Jamie Carnie is an independent philosopher and software engineer. He worked on one of the UK's first popular philosophy magazines and his re~~...~~ ~~...~~l on perception and consciousne~~...~~

Blue Sky
THOUGHTS

Colour, Consciousness and Reality

Published in Great Britain and the United States in 2007 by

MARION BOYARS PUBLISHERS LTD
24 Lacy Road, London SW15 1NL

www.marionboyars.co.uk

Distributed in Australia and New Zealand by
Peribo Pty Ltd, 58 Beaumont Road, Kuring-gai, NSW 2080

Printed in 2007
10 9 8 7 6 5 4 3 2 1

A CIP catalogue record for this book is available from the British Library.
A CIP catalog record for this book is available from the Library of Congress.

ISBN 0-7145-3124-3
13 digit ISBN 978-0-7145-3124-3

Set in Bembo 11pt

Printed in England by Creative Print and Design

Cover design by Holly Macdonald

Blue Sky
THOUGHTS

Colour, Consciousness and Reality

by Jamie Carnie

MARION BOYARS
LONDON • NEW YORK

for Fiona

Contents

PART IV: CONSCIOUSNESS (SOLVING THE HARD PROBLEM)

PART V: CONSEQUENCES

***

It is possible to say something true. For example: 'something exists'. The truth of this statement cannot be doubted. It is uniquely certain because it is unquestionable that we as humans experience. It would be impossible for experience to occur and yet for absolutely nothing to exist.

But if we know with complete certainty that something exists the next question to be asked is, what is the nature of this existent 'something', or reality? Given that its existence is implied by experience it is likely that an answer to that question is best approached by ascertaining the nature of experience itself.

***

INTRODUCTION

Colour appears all around us. From the blue of the sky to the green which dominates the natural world, and the bewildering diversity of hues displayed by flowers, fruits and animals. What is common to all occurrences of colour, however, is that each individual instance is extremely simple. The field of red visual quality visible on the exterior of tomatoes, for example, does not contain any distinguishable elements or structure. One consequence of such simplicity is that the contrast between the appearances of different hues – red, green, blue, yellow and so on – is maximised, and the chromatic variety offered to us by the surfaces of the world is as great as it could possibly be. Another consequence is that colour provides one of the most basic forms of content contributing to the appearance of the world. The other forms of content that make up the appearance of the world are the remaining sensory qualities of sound, smell, taste and heat. These too are equally simple in nature. Thus individual occurrences of the aroma of baking bread or the sound of a firework exploding are – along with colour – as simple entities as one will ever encounter in the universe.

Yet despite their significance as providers of content, and their ubiquity, man has yet to achieve an understanding of the nature of the sensory qualities. This is remarkable given their simplicity. They hardly present the technical challenges of such ultra-complex objects as brains or galaxies, which contain billions of components (nerve cells and stars respectively). It is all the more extraordinary, too, in the present age, when scientists have developed their disciplines to the point of being able to explain events in the physical world at scales of structure ranging from the interior of the atom to the entirety of the cosmos. The position is somewhat analogous to that which notoriously obtains in the field of medicine regarding the 'common cold'. Just as this most widespread and down-to-earth of afflictions remains incurable while treatments have been found for more esoteric diseases, so here the simplest features of the physical world, and those which have the most direct impact on man, remain a mystery while nature's most complex structures are steadily being laid bare.

Of course, science has revealed a great deal about the events that usually occur in the material realm when sensory qualities become evident. Physics has told us about the nature of electromagnetic radiation ('light' as we call it when it vibrates at frequencies that our eyes can detect), which reflects off objects when they appear coloured, and radiates from them when they feel hot. It has also told us of sound waves that in a similar way are associated with the quality of sound. Chemistry has revealed the make-up of molecules which, when they diffuse through air, give rise to smells. But there is no prospect of the disciplines of science shedding any light on the mysterious nature of sensory qualities themselves – the blueness that we see in the sky, the sound of leaves heard rustling in the wind, or the warmth felt as given off by a crackling fire. This is precisely because of their simplicity and content-providing nature. For these characteristics make the sensory qualities quite unlike the types of entities that science is capable of dealing with.

In contrast to, say, atoms, which are made up of an internal structure of component particles such as electrons and protons, entities like the colour red and the smell of thyme possess no structure and hence have no measurable or quantitative features. As they are neither quantitative or measurable they cannot become subject to the mathematically-based (therefore quantitative) forms of theorising that prevail in science. For the same reason no instrument could ever be built that was able to detect the sensory qualities (as distinct from the signals such as light or sound waves, which are associated with them) so they could never be investigated through science's experimental procedures. In short, the sensory qualities are invisible to everything in the scientist's tool-kit. The combination of super-simplicity and content-provision that characterises the sensory qualities – often referred to by saying that colours, sounds and so on are 'qualitative' rather than quantitative – enables them to slip through science's otherwise all-conquering ability to tease out the structures of reality.

If questions about the sensory qualities cannot be addressed by science then the techniques of philosophy may offer a more promising alternative. If and when philosophers ever manage to answer them, we can expect the answers to have implications both in the arena of the 'external' physical realm and in that of the 'internal' mental one. This is because perception, the activity in which awareness of sensory qualities invariably arises, acts as a bridge between these two domains. Inevitably, therefore, any consequences which will follow from our coming to understand the nature of the most basic entities to arise in perception, will fall on both sides of that bridge.

But it has to be said that advances in achieving a philosophical understanding of the nature of the sensory qualities have so far occurred at a less than impressive rate. In one form or another, the question of the nature of the sensory qualities has formed a topic of consideration by thinkers (in East and West) for many centuries. Yet even now there remains little or no sense of movement towards a complete solution. Why is progress so restricted? The reason for this, I suggest, is the same as the explanation for philosophy's poor performance in answering many of the questions which confront it. The method used to create solutions to philosophical problems is non-repeatable. In consequence, the discipline is arguably still in its Stone Age.

If my experience is anything to go by, what philosophers do when confronted by a problem is to 'sense out' a solution to it. This is a lengthy and quasi-artistic procedure which is similar to how a painter or sculptor forms their work. It is also, no doubt, akin to the semi-intuitive procedure that would have been acquired, after years of practice, by Stone Age 'knappers' – who were, after all, the sculptors of their day – in order to strike a piece of flint in the correct way to fashion neolithic implements of utility.

In philosophy each new problem needs to be 'sensed out' from scratch, just as a Stone Age knapper faced with a fresh piece of raw flint would first have had to feel it all over before creatively applying his intuitive skills and working out at what angle and force to strike it. Because the methodology is intuitive and creative, for both the Stone Age worker and the philosopher, each new problem demands an approach which, although it may build on skills acquired in previous attempts, is unique to the particular problem being faced and is never repeated for others.

This is all quite different from what happens in science, which through its use of mathematics for the development of theorems and a standardised protocol for experimentation, is in possession of repeatable procedures that are applied again and again to every new problem that comes up. No-one would deny that there is also a significant element of creativity in science, but the foundation of its methodology – in contrast to that of philosophy – involves a set of procedures that are infinitely re-applicable to any problem that may be confronted. In this respect, extending to it the 'pre-historical' metaphor applied to philosophy, one may say that science is in its Bronze Age. Bronze Age metalworkers discovered how to combine a standardised 'recipe' of ingredients under heat to form bronze, and thereafter were able to repeat the same operation at will in order to produce metal ingots to order, just as scientists can repeat their standardised recipe of procedures at will in order to solve virtually any problem that may arise.

People react to the primeval position in which philosophy finds itself in different ways.

Amongst the tribe of philosophers one notices, for example, endeavours to emulate some of the superficial features of science's success, such as attempts to apply methodologies which have a semi-scientific ring to them, or the finishing of texts to a high level of technical language. Reactions such as these are somewhat like members of the unruly tribe of philosophers jealously watching the metal swords and jewellery of the advanced tribe of the 'Scientii' in the neighbouring valley glitter in the sun, and taking to polishing their crude stone artefacts in a vain effort to achieve a similar effect.

Another possible reaction to the condition of philosophy could be to think that if the discipline is perhaps only in its Stone Age then realistically it stands little chance of answering any of the questions which confront it. But I would commend an alternative attitude to this which offers a more positive way of looking at the situation. If the possibility exists that philosophy is currently only in its Stone Age, then there may as yet be great tracts of discovery to be made and huge fields of development for the discipline to undergo in the future. After all, once it develops beyond its Stone Age, philosophy will presumably be as different from the subject we know today as chemistry currently is from alchemy. So on this view the possibilities offered by the present situation are considerable and there is a significant chance that there may genuinely be answers 'out there' to many of philosophy's questions.

In this book it is this positive outlook, and the associated attitude that philosophical questions are in principle answerable, that I have attempted to bring to bear on the problem of the nature of the sensory qualities. In trying to find that answer I have endeavoured to strike a blow at the 'flint' of the problem at an angle that to the best of my knowledge has not been attempted before. Whether or not it succeeds in revealing a form of answer that carries any worth will be for others to judge.

PART I:
COLOUR
UNCONFINED

1. Blue Sky Thoughts

Imagine a beach. The steep path which you have been following over an arid hill top from the nearby fishing village, after a pleasant midday meal and a glass of wine at the local taverna, has reached the end of its course through parched shrubland that is richly scented by mint, laurel and myrtle, and sparsely broken by dried-out trees. From this elevated vantage point the shape and setting of the beach are now fully visible. It is a great golden arc of sand along the curve of a wide Mediterranean bay which separates grass-fringed dunes from an open vista of glistening sea, where waves roll in rhythmically and curl and break in the shallows. Along the belt of sand numerous bathers are scattered in groups, taking advantage of the powerful sun which beats down from an unbroken sky of inky, cerulean blue above. Two beach cafés nestled close in by the dunes are attracting custom by offering a refuge under their slatted bamboo roofs from the relentless heat, and a ready supply of cold drinks.

Upon leaving the slight shade afforded by the trees you instantly feel the full intensity of the heat. The sun burns strongly on your arms and face and, when you descend to the beach and remove your shoes, you find that the sand is almost painfully hot to walk on. But a neat line of indigo blue sun-loungers, each with its own pale turquoise sun-shade, forms a laddered gash of alternating colour and shade along the centre of the beach. Having checked in your pocket for the right number of euros needed to hire one, you walk across to the nearest and settle down on it, pulling out a thriller to read from your bag.

After twenty minutes of dozing and reading in the sun the attractions of indolence begin to wane and you sit up to examine your surroundings. The first thing that catches your interest is the sea and the sunlight which sparkles off its many wave crests like an army of distant flashlights. But the effect is tiring on your eyes, so soon you let your attention drift skywards to the visually more relaxing vista of boundless blue.

DEEP BLUE

During the next hour the peaceful sky becomes an ever more intriguing object of contemplation as the idea slowly develops that there is a unique character to its colour. This realisation is brought about by a subtle contrast between the coloured appearance of your sun shade and that of the sky. Looking directly upwards your field of vision is partially filled by a segment of the underside of the parasol. The sun-bleached turquoise of the umbrella's material is slightly darkened by its own shadow and as a result its hue perfectly matches that of the sky, resulting in a single glorious expanse of blueness across the entirety of your visual field. Yet despite their identity of hue there is a marked difference between the appearances of the two visible regions of blue colour. That on the underside of the umbrella seems flat and surface-like, whereas that in the sky looks as though it is radiating downwards from a region high above in the atmosphere. The blueness of the sky has a quality of *downward-pointing* to it, and appears to reach out from its location in the sky, whereas the blue of the umbrella looks as if it is simply attached flat on the surface of the material like a two-dimensional field of colour that is integral to and wholly contained by the material.

This sets you to thinking about colour more generally. It is evident that most colours visible in the terrestrial environment have the appearance of two-dimensional or 'flat' surfaces more akin to the blue of the umbrella than that of the sky, because they reside – or at least appear to reside – on the exteriors of physical objects. At this point you look around seeking further examples. The red colour on the Stella Artois beer signs outside the beach cafés offers one. Another, you notice, is the lurid purple of the plastic plates being used by a nearby French family for their picnic. The appearance of these colours could be likened to fine coatings of redness and purpleness respectively on the external surfaces of the physical objects. But the azure of the sky is unlike such flat colours. It does not look like a blue coating on any kind of a surface high in the atmosphere. When you look upwards at the sky it doesn't seem as though there is a surface up there on which the blueness might reside. And more than that, the blueness appears to project downwards from its location in the atmosphere in a way that no natural surface colour ever does. It is almost as if the sky is a great field of blue that is glowing downwards towards the earth. Furthermore there is, of course, no solid, physical surface in the sky to which the blueness could belong. So if the blueness of the sky is not a normal surface colour what is it? Could it be, the thought occurs, more like a translucent one which belongs to the entire volume of the atmosphere? After all, the colour of the sky is caused by the scattering of sunlight from the air molecules which make up the atmosphere,[1] and there are many other such translucent colours in the natural world; colours that don't occur merely on surfaces, but permeate through the volume of things.

At this moment you notice some refreshments which the father of the French family has prepared to accompany their picnic. In two glass tumblers he has poured first some *cassis* and then generous measures of white wine. The translucent, deep red colour of the resulting drinks of *kir* is not on the surface of each glass but extends throughout the volume of the liquid. If the blue of the sky is in the same way a feature of the entire volume of the atmosphere this would explain why no surface of blue is to be found when one looks up into the sky. It would also account for the sense of depth that one experiences in its appearance. But there remains still the fact that the colour of the sky manifests downwardly from its location. This would not in itself be explained by the possibility that the sky's colour permeates the volume of the atmosphere like a translucent colour.

SPHERE OF LIGHT

You consider the enormous panel of unblemished blueness above as you absorb these ideas. Doing so your gaze briefly sweeps over the sun, now somewhat past its zenith. At that moment a new thought occurs. If the blue of the sky is not a surface colour then certainly the colour of the sun cannot be either. The sun is a distant globe of seething hot gas which we experience as yellow in colour. Yet its yellowness, just like the blue of the sky, is quite unlike any normal surface colour. This thought is reinforced when you notice two boys playing football. They are doing so with a bright, saffron yellow, plastic beach ball. If the sun's yellow were like a 'normal' surface form of colour this, you realise, is how it would appear. The solar globe would look something like a giant yellow beach ball hanging in distant space. (Perhaps somewhat like a brighter version of the moon.) Yet in actuality the sun appears nothing like a yellow-coated sphere at all. Its yellow colour literally blazes out towards us from the solar body.

You lie back on the sun-lounger to assimilate further these unfamiliar new concepts. As you consider them it gradually becomes evident that a common feature distinguishes the colours of the sun and sky from other instances of colour. In both cases the differentiating factor is that they have the appearance of not being confined to their physical locations. Thus while surface and translucent colours appear to reside in a simple fashion on the surfaces or in the volumes defined by their 'host' objects, what is special about the blue of the sky and the yellow of the sun is that the same is not true. They appear to manifest outwardly from their source locations. The blueness of the sky does so downwardly from the atmosphere, and the yellowness of the sun does so outwardly through space from the substance of the solar globe.

BUBBLE OF SOUND

A few moments later there is a sudden and fleeting instant of coolness. You open your eyes to see that a tall man has just walked by. His shadow has passed over you and interrupted the warming rays of the sun. This brings to mind the fact that people experience not just the colour of the sun but also its heat. As you lie on the lounger feeling the intense rays rapidly warming you back up you mentally compare the form displayed by the sun's qualities of heat and of colour. The sun's quality of heat, you realise, is no more contained on its surface than is its colour. Just as colour is manifested outwardly from the solar globe so is heat. The sun's quality of heat reaches out across space to be felt here on the beach, as does its colour.

You shut your eyes, intending to consider such parallels further, but a short while later the light sleep into which you unintentionally fall is interrupted by a gentle buzzing sound. It seems to be coming from the far end of the beach. Sitting up you scan the sky in that direction and eventually make out a small green triangle which is just visible in the distance, drifting towards you, high above the dunes. As the shape approaches over the next few minutes the volume of its sound gradually increases and the triangle soon resolves into the form of a microlight aircraft. The pilot circles the craft lazily over the beach a few times and while it is overhead you find yourself considering the noise that it produces. The high-pitched drone made by its straining two-stoke engine and the propeller biting into the air is not confined to the region of the craft. It does not give the appearance of there being a bubble of sound locked into the immediate space around the craft which travels along with it as it flies. Rather, the sound quality reaches out from the craft towards the surroundings in much the same way as the colour of the sky and of the sun manifest outwardly from them.

Sound it seems, displays the same characteristic of extending outwardly as colour and heat do. In fact, you note, in none of the cases considered so far has a quality been found to give the appearance of manifesting intrinsically within its object. Thus, neither the colour nor the heat of the sun appeared confined to its outer layer. The sound of an aircraft did not arise as a bubble of auditory quality around it, and the blue of the sky was not fixed on a mythical surface floating high in the atmosphere. Rather, all of these qualities have been found to manifest outwardly from their apparent locations in the physical world.

A little later you decide to take a stroll. You head towards the dunes and then along the back of the beach and across to one of the beach cafés. Here a chef has laid out the first of the day's sardines on a steel rack above charcoal which glows red-hot, and has started grilling them along with a selection of vegetables. Together with a small crowd which has gathered at the bar you watch the process for a few minutes.

As you observe the sizzling ingredients you can't help noticing that their aroma permeates outwards from the grill-rack to fill the space around the café. No matter what position along the bar you adopt as vantage point the same mouth-watering smell reaches out to it. It occurs to you that odour has the same characteristic of manifesting outwardly that colour, sound and heat do.

NOT SO SIMPLE

It seems natural to think of the elements of sensory experience – the sensory qualities of colour, sound, smell, heat and so on – as integral to and wholly confined within regions of the physical realm. But the observations here on the beach indicate that each of the sensory qualities manifests a degree of extending outwardly. This shows that any conception of the sensory qualities as limited to restricted regions of the world misses out a key feature of their nature. For a number of centuries the dominant intellectual view has been that sensory qualities are mental constructs, but even here the idea that sensory qualities do not extend outwardly has prevailed. This is exemplified in the tendency acquired in the modern world (perhaps because of its preponderance of man-made surfaces) for thinkers to characterise such mental sensory qualities as two-dimensional in nature. Thus at the opening of the 20th century leading British philosopher Bertrand Russell went so far as to account for sensory experience as being composed entirely of 'patches' of such elements.[2] In more recent times the respected neurologist Antonio Damasio used the more sophisticated but nevertheless still 'flat' metaphor of a cinema screen when he posited that the sensing mind is like a 'movie' in the brain. He claimed that: 'Qualia are the simple sensory qualities to be found in the blueness of the sky or the tone of sound produced by a cello, and the fundamental components of the images in the movie metaphor are thus made of qualia.'[3]

For both of these leading thinkers, then, the essential feature of human sensing which has been identified here on the beach is to be discounted. What man's perceptual experience amounts to is simply a set of flat regions of sensory quality which come together as a shifting array in the forum of the mind, somewhat like the elements of a dynamic mosaic. From this array the mind formulates a picture of the three-dimensional world, much as the illusion of a view with depth can be derived from the two-dimensional areas of colour on a photograph or cinema screen. But the examples identified on the beach show such ideas to be overly simplistic. They indicate that sensory experience contains subtle features that crude pictures such as those of Russell and Damasio cannot account for. For on the beach we have seen that the sensory qualities of colour, sound, smell and heat all display a common characteristic of manifesting outwardly from their physical locations. Furthermore, another plausible conclusion that can be linked to our observations is that outward manifestation of sensory qualities is always associated with an appearance of directionality. Thus the blueness of the sky manifests itself in a downward direction, but it never, for example, manifests itself sideways across the atmosphere, nor upwards into space. Likewise, when observed from a distance the sound of a microlight emanates outwards from the aircraft but never inwards back to the craft. Apparent directionality and outward manifestation seem to be closely combined in most, and perhaps all, sensory qualities.

That there may be these twin characteristics universally amongst all sensory qualities can hardly be an insignificant matter. Could it be that there is more to understanding the sensory qualities than has been appreciated so far? In this book I will endeavour to explain that this is indeed the case. But more than that, I will argue that the difficulties presented by one of the hardest problems man currently faces in trying to understand the nature of consciousness – the question of how it arises from the physically nondescript grey-matter of the brain – are due to oversimplification of the nature of such sensory elements, in the manner of thinkers like Russell and Damasio. When the full and subtle nature of the sensory elements – as revealed in the blueness of the sky and other such phenomena – is properly taken into account, this allows the jigsaw of consciousness and its relation to the physical realm to be pieced together in a remarkably neat and altogether revolutionary fashion.

The issue of outward manifestation is not the only one concerning the sensory qualities that it will be necessary for us to examine in order to fully explore the nature of sensory consciousness, its relation to the physical matter of the brain and to the physical realm more generally. In addition we will have to address what is perhaps one of the most profound questions about the sensory qualities of all.

2. Forest of the Senses

There is an old question, often attributed to Zen Buddhism, about whether a tree falling unheard in a remote forest makes a sound…

Imagine, therefore, a great wood of evergreens in the foothills of a high mountain range many kilometres from the nearest human habitation. In the depths of this endless forest one pine tree stands out taller than most. The tree is ancient and the time has come for it to yield its place to a new generation of saplings which are thrusting up into every available space in the tightly packed woodland. The tree's great height is further accentuated because it rises from the crest of a ridge between two valleys.

The roots of the upstanding tree are old and decayed and they have been tested beyond endurance by an overnight gale which sent waves of turmoil through the wood like a field of wheat being ruffled in a breeze. Standing in its prominent position where it was exposed more than most to the tugging of the wind, the majestic pine now sways backwards and forwards as if undecided about its future. Then, a sudden slumping motion jolts it as its deepest roots give way, and at last the grand old pine pitches into space. The huge trunk arcs forward smoothly, while at its base its network of roots are snapped in quick-fire succession and tear a great mound of earth from the surrounding soil. The descending trunk smashes effortlessly through the foliage of nearby trees, snapping off even the thickest of their branches as if they were twigs. Upon impact with the ground a great judder is sent through the length of the trunk, causing it to rebound momentarily into the air, and vibrations emanate through the surrounding soil making the nearby vegetation shake in a wave of frantic trembling.

Yet the nearest human being to this crescendo of events is a farmer patiently tilling his fields beyond the edge of the forest a number of kilometres away. So it all plays out far beyond the range of human ears. The question which Zen Buddhism challenges us to answer is this: do such events occur with or without a quality of sound? Does the storm of physical events which make up the fall of the tree produce an apparent noise – despite being unheard – or does it all occur in utter silence, as in a silent movie?

Yet such a forest, so far from human beings, can also be the setting for many similar questions about the other sensory qualities. For example, consider a small clearing at the bottom of one of the valleys beyond the ridge where the tree has fallen. Here, amongst the berries, flowers and stones are occurrences of each of the other sensory qualities, all equally hidden from human awareness. For example, nestled amongst the undergrowth which makes up one side of the clearing, is a wild rose which is producing its first flowers of spring. The young petals reflect from the sunlight that is falling upon them a wavelength of light which we would call 'pink'. But unseen by any human eye do the petals display this – or any – colour quality? The flowers are also emitting a fine gas of complex organic molecules, but in the absence of a human to detect their scent does this mean that the flowers are presenting a perfume to the world around them? Further along is a thicket of brambles laden with juicy blackberries that are ripening in the midday sun. Yet without being placed in a person's mouth have they any sweetness? What about the bare outcrop of rock which protrudes from grass in the middle of the clearing? It has had the sun shining down upon it from overhead all morning.

To the human touch the rock would feel, if not hot, then at least distinctly warm. But there are no humans here to touch it. Despite this, does the rock somehow present a quality of warmth to the nearby space?

THE SECRET LIFE OF THE SENSES

Along with the fallen pine tree, the questions posed by the clearing in the 'forest of the senses' illustrate a common point, which is that all of the sensory qualities are open to the same query: do they exist in the physical world in the absence of human perception? This is partly a straightforward question about whether such qualities can exist in a domain that is known to be made up of objective particles such as atoms and molecules. But also partly a not-so-straightforward question about how, if they do, they can make themselves manifest to an observer-free environment – is it possible for them to have some sort of 'secret life of appearance' all of their own?

A good way to get a picture of what this latter question means is to consider it when extended to an extreme situation. And what could be more extreme in terms of lacking observers than outer-space? Consider, therefore, the following extra-terrestrial example of sensory qualities without perceivers. It is offered by Beagle 2, the European Space Agency's lander on Mars, which in 2003 attempted a descent to the planet's surface from the 'Mars Express' spacecraft. Unfortunately, the likelihood is that the probe never made it intact to the surface because no data was ever returned from the vehicle once on the planet. However, its fate was not established with certainty and there remains the slim possibility that it may have endured the descent and subsequently experienced a communications fault once on Mars. This means that despite the expedition being a disaster from a scientific point of view, it may – if the vehicle survived – have unintentionally given rise to a philosophical triumph. For the Beagle 2 Lander had a unique feature. Bolted to its side was a small and colourful piece of artwork on a triangular slab of metal by the artist Damien Hirst.[1] It was one of Hirst's 'spot paintings', composed of dots of colour on a plain background (albeit at a size which could fit in the palm of a human hand, considerably smaller than was usual for his gallery pieces). It was not just an aesthetic gesture on the part of the Beagle team to invite Hirst to contribute a spot painting, for the work was to be used to help calibrate on board cameras and spectrometers. But whatever their motivation, if the vehicle did manage to survive its undoubtedly tricky landing, then for a period of time there may have existed in the lonely isolation of the boulder-strewn landscape of Mars (and perhaps even now still does) a small metal triangle covered in dots of differently coloured pigment which was bolted to the side of a battered and dust-coated little machine.

Mars is not only a place where no humans have ever been, scientists do not even know if there is life of any sort on the planet. So with Beagle 2 and Hirst's spot painting we get the question of the secret life of colour in an historically extreme form. For if sensory qualities such as colour do exist in the physical realm without any dependence on human perception, then it may be the case that even now at a lonely spot some millions of kilometres from earth in the barren terrain of Mars the red, orange, green and blue colour qualities of Hirst's spot painting are doing so. In which case they are manifesting as appearances to...well, to *what*? The answer cannot be that they are manifesting to any form of consciousness, because there is none on the planet of Mars. So it can only be that if the colours are manifesting at all they are doing so to the surrounding, entirely non-sentient locality. The colour qualities of Hirst's little spot painting are appearing, if at all, to an environment which is devoid of even the slightest consciousness. More generally, it is a corollary of the idea that sensory qualities exist independently in the physical domain that they can manifest to environments that are entirely free of observers. There are many other examples of locations in the universe which illustrate the truth of this point, beyond the case of Beagle 2. From deep in unexplored ice crevasses here on earth to lava flows on the moons of Saturn, there are a multitude of locations in the cosmos which on the view that sensory qualities exist independently would have colour quality present but at any given moment are entirely free of sentient observers.

THE NATURAL VIEW

To return to the clearing in the 'forest of the senses'. There are only two possible answers to the question of whether sensory qualities exist in the physical world without being perceived: either they do or they do not. Let us examine the implications of each of these two responses.

According to the first answer the way the world 'looks' (and sounds, smells and tastes) when we are not observing it is exactly the way that it does when we are. In the case of colour this can only be because the fields of visual quality which we call 'colour' are genuinely located on the surfaces (or in the volumes) of physical objects. In the case of the other sensory qualities it can only be because the regions of auditory and olfactory quality which we call 'sound' and 'smell' are actually located 'out there' in the spatial environment just as they appear to be. On this picture, a wild rose which is hidden from human eyes because of its location deep in a forest, is nevertheless pink because its colour quality is actually on (or perhaps in) the flesh of its petals. Equally, despite the lack of conscious beings on Mars, Hirst's spot painting attached to Beagle 2 does, somewhat in vain, display its aesthetic glories of red, blue, green and so on because these colour qualities are actually on the surface of his painting. Likewise, a tree which falls in a remote forest makes a 'sound' because the auditory quality which it generates is physically located in the surrounding space. Anyone wishing to accept this view has to be prepared to accept also, as we have seen, the somewhat unfamiliar position that such non-dependent sensory qualities manifest to the observer-free environment (in the sense of simply the inert space of the surrounding physical world).

But the second answer to the question posed by the forest of the senses leads to a very different view. Proponents of this alternative way of thinking maintain that sensory qualities can have no existence independent of our perception of them. The simplest and most obvious way in which this might be the case is if they are mental entities. There are good grounds for thinking this to be so. Colours, sounds, smells and other sensory qualities share a number of characteristics, such as simplicity and experiential directness, with bodily sensations like pain and tickles which are widely held to be mental. Also, of course, we can experience colours and the other sensory qualities, or at least residual versions of them, in purely mental arenas such as those of the imagination and dreams. According to such a picture, then, all of the physical events in the forest of the senses that we have been considering are unaccompanied by sensory qualities because there are no perceivers within the forest to give rise to them. On this interpretation, the forest becomes literally a realm without appearance. Likewise the spot-painting on Beagle 2 consists only of a block of metal with some daubs of pigmented chemicals. It displays no colour qualities to the Martian environment because none exist intrinsically on its surface.

It would be wrong to claim that anyone thinks the truth of either of these two pictures is a matter of common-sense, but it is the case that the idea of sensory qualities as existing independently and objectively within the physical arena is the one that most people (of all cultures) actually live according to and take for granted in their daily lives. Thus for example, at the simplest level, we all believe that, and behave as if, the colour qualities of things persist when we shut our eyes. Likewise, when an item such as a can of baked beans is placed out of sight in a well-lit cupboard no-one questions that its label retains its famous '57 varieties' blue colour. Few of us even doubt that the beans inside the can, sealed off as they are from all light, retain their orange colour. In a similar way, most people would consider that, despite being unheard, a china plate falling to the floor in their house while they were out would produce a quality of sound – and a loud one – when it shattered to pieces. Our beliefs concerning the sensory quality of heat are of a similar nature, as shown by our attitudes to the heating elements of electric fires and cookers. For example, most people consider that when switched on these continue to give off an intense quality of heat, along with a dangerous amount of energy, even when they are not actively being felt.

The package of ideas centred on the view that the sensory qualities exist 'out there', independently of man, agrees so powerfully with the way that things appear in the world that it seems in many respects as though it 'must obviously' be correct. For this reason – and also because it accords with the practice of most peoples' lives – I will call it the 'natural view' of the sensory qualities.

Natural Doubts

Yet despite taking the natural view for granted in daily life, few people would be prepared to state categorically that it represents the final truth concerning colour, sound, smell, taste and heat. Why is this? It is because there are serious doubts concerning the viewpoint which, at one level or another, most of us have some awareness of. In many cases these stem from the difficulty of assimilating the natural view to the picture which science has given us of the physical realm. To take an example: the scientific explanation of colour is in terms of light being reflected off surfaces and into the eyes of the viewer. This causes electro-chemical signals to be transmitted up the optic nerve. Colour arises when the visual centres of the brain process the resulting information. Likewise, sound is produced when sound waves stimulate fine hairs in the inner ear which in turn generate signals that are sent into the brain via the auditory nerve. But what happens in a situation – such as that of Beagle 2 or the tree falling in a remote forest – where there is no perceiver, and thus no signals are being processed in a brain?

The scientific picture tells us that in these circumstances there can be no quality of colour or sound because the mechanism on which they depend – the brain and perceptual system of the perceiver – are not present. But this is in direct contradiction with the natural view that incorporates the idea that sensory qualities are manifest even in the absence of perceivers. So, on the central question of whether sensory qualities can occur in the absence of perceivers, the two accounts lead to mutually conflicting positions. But the natural view also has difficulties in relation to the scientific picture on a second front. For, as we have seen, it is part of the natural view's 'package' that perceiver-independent sensory qualities can manifest to environments that contain no observers. Again, here science tells us about what is going on physically when such events occur. Thus when the colour qualities of Hirst's spot painting on Beagle 2 'present' to the inanimate world of Mars – if indeed they do – light from the sun is reflected from the pigments of the painting's surface and then spreads out through the alien atmosphere.

But is it an adequate explanation for the manifestation of a colour into its surroundings solely that a physical signal like light travels to the point where the appearance occurs? Equally, can the manifestation of the quality of sound at a point just be a matter of sound waves having spread out into the nearby environment from an event such as a tree falling or a plate crashing? It seems as though there is some vital ingredient missing from such explanations, which is required in order to account for the essential nature of sensory qualities as appearances.

Then what about the question of atoms and molecules? When a person says that the surface of an object such as that of a car is red is he or she claiming that the paint molecules in which it is coated are imbued with the quality of redness? If so, is the car red because each molecule and perhaps even each individual atom is infused with the quality of redness? But given what is known about atoms – not to mention even smaller particles such as quarks[2] of which the same question might also be asked – this seems inconceivable. Even at the larger scale of the everyday there are difficulties enough about defining the precise location of so-called 'surface colours' in a world where science has taught us that few structures are as simple as they look to the naked eye.

Consider, for example, the colourings of flowers. Wild roses such as those in the clearing in the forest of the senses whose colour we contemplated earlier have petals of a fine and flimsy nature. Yet even with them there can be ambiguity concerning whether their pink colour is *on* the fabric of the petal or *within* it. The issue becomes even more pronounced with flowers such as tulips which possess distinctly thick and fleshy petals. When these are dissected it can be seen under a magnifying glass that they have a complex, multi-layered structure, and when viewed through a microscope their cellular complexities become more fully evident. If it were claimed that a tulip's red colour is located on, or even in, the body of its petals then the awkward question would arise of precisely where in relation to these strata of cells that position is? At the atomic level the challenges posed by such questions become infinitely greater.

There would, then, be major problems in any attempt to embed the natural view in a science-based understanding of the physical realm. Defining the physical location of the fields of visual quality that we call 'colour' with respect to even the fabric of the physical world that is visible to the naked eye, or the eye aided by the microscope, is difficult enough. When it comes to doing so relative to the forms of matter – such as atoms and quarks – that man has been made aware of by his most advanced experimental procedures and scientific theories the task appears impossible. But worse than that, the natural view also leads to a paradox, for if consistently applied it would imply that opaque objects such as bricks and stones have interior colours. This is because it is part of the 'package' that colour qualities don't need to be experienced in order to exist, so the fact that we can't see inside these bodies does not preclude them from having such colour qualities. Their interiors are made of the same material as their exteriors, the only difference being that light cannot reach the interior. Yet black, the colour we experience when no light is reflected from a surface, is one of the visual colour qualities just as much as any of the others – it is to be found on the surface of coal in the same way that yellow occurs on the surface of lemons or green on grass, and in general black occurs in the visual realm on an equal footing to other experienced colour qualities – so could it be that the interiors of opaque bodies are black? Possibly so, but this clashes with our practical conceptions of things. Who goes around thinking of the inside of stones as black? On the contrary, most of us assume that the interior appearance of objects is continuous with their exterior appearance. We noted this earlier in the case of the baked beans sealed up in their can. Despite the lack of light, few people think of the beans in their enclosed can as anything other than orange in colour. It would require a special effort of thought to think of them as being black.

It is at this point that one realises that as a day-to-day basis for living our everyday thoughts about colour may be all very well but they lack certain theoretical niceties such as consistency and completeness. It is at least in part because the natural view lacks such consistency that it fails to sit comfortably with the scientific outlook which is *par excellence* man's most thoroughly developed body of systematic knowledge.

Ultimately, however, the main source of doubt about the natural view for many people lies not so much in these technical philosophical points, significant though they may be. Rather, it stems from the existence of an alternative and highly successful way of explaining perception. This is an account which is closely tied into our most fundamental scientific concepts, and which therefore suffers none of the difficulties of the natural view in relating to the scientific picture of the world. The rival explanation of colour and other sensory qualities has turned out to be so powerful in its ability to explain perception that it has sown widespread seeds of doubt over the truth of the natural view. Not only that, but because of its ability to account for various aspects of perception which the natural view has difficulties with, it has grown to become the dominant account of perception amongst both scientists and philosophers. In order to examine it we will start by considering its historical origins.

3. Descartes' Perception Machine

The discovery of an alternative to the natural view was a two-stage process. Firstly came the development of an explanatory model for perception that acknowledged its dependence on entirely physical processes inside the brain. This was followed by an enhancement of the initial picture, which applied to it a deeper understanding of the nature of light and its relation to colour. These steps were carried out by two of the great thinkers of post-Classical Europe: the first by Descartes (1596-1650) and the second by Newton (1643-1727).

The French philosopher and mathematician René Descartes is best known today as the author of the famous phrase '*cogito ergo sum*' ('I think, therefore I am').[1] This quote suggests a picture of him as a person who was remote from the everyday world and interested only in issues of abstract metaphysics. But in fact there were other, highly practical, dimensions to his work. He lived during the 17th century when the scientific revolution which laid the foundations of the view of the physical world that we take for granted today was gathering pace. Ancient Greek ideas about matter being made up of large numbers of indivisible 'atoms' had been rediscovered and were being championed by thinkers at the cutting-edge of the new intellectual developments. Amongst this group was Descartes himself who took up the novel atomistic way of thinking enthusiastically and contributed in a number of ways to its eventual entrenchment. Not least of these was by demonstrating the possibility of constructing an entire cosmological theory based upon it. Descartes also made ground-breaking contributions to the new technique of experimental science. These were mostly in the nascent field of optical science and included the discovery – or at least the first published mathematical description of – the law of refraction.[2] From this he was able to calculate how to create lenses that produced clearly focused images. Also, to work out how the human eye operated, and to explain the colours of the rainbow by observing the reflection and refraction of sunlight in a glass flask filled with water.

It is to the experimental and philosophical work carried out during the 17th century – in part by Descartes – that we owe the consensus that exists today that matter is made up of particles such as atoms. But the almost equally strong agreement that the natural view does *not* represent the last word regarding the sensory qualities also derives from the work of Descartes during this period.

MONSIEUR DESCARTES DISSECTS

Associated with the atomistic outlook on matter was a mechanistic form of explanation of physical phenomena. (This was part of the outlook's appeal in an age when simple mechanical devices driven by water-power or clockwork represented the pinnacle of technological achievement). Thus the atoms or 'corpuscles' were conceived of as little granules of matter in possession of shape and size. Leading thinkers of the time, such as Dutch scientist Isaac Beeckman, explained phenomena such as wetness, for example, as resulting from bodies being composed of 'sharp' atoms.[3] Similarly, Descartes' own cosmology was built around a mechanistic version of nucleosynthesis in which large numbers of tightly-packed corpuscles undergoing rotary motion in the pristine conditions of the early cosmos ground each other down into new shapes and sizes which gave rise to the various forms of matter currently observable in the universe. One of Descartes' subsequent insights was to work out how to apply this *mechanical* form of explanation to the problem of perception.

Descartes was a thinker of almost unbounded vision. Having created an entire cosmological theory out of the unpromising raw material of 'corpuscles' it would hardly be an exaggeration to describe him as the Einstein of the Mechanical Age. But not content with constructing a corpuscular cosmology, he eventually conceived the possibility of extending the mechanistic form of explanation to the whole of reality (to 'The World' as his subsequent *magnum opus* in the field would, appropriately, be titled). But this meant incorporating man into the picture as well. In order to do this Descartes recognised that he would have to gain a complete understanding of the physiology of the human body, and – because the body's internal structures were ill-understood at the time – this could only be achieved by taking up the study of anatomy. So he threw himself into the task of acquiring anatomical knowledge with considerable vigour, dissecting cattle parts obtained from a local butcher, and achieving such skills in anatomy that by the 1630s, whilst living in Amsterdam, he was invited to assist in the demonstration dissection of a human cadaver at an amphitheatre in the city.[4] Indeed his knowledge of anatomy became so highly regarded that one recent commentator has suggested that if he were alive today Descartes 'would be in charge of the CAT and PET scan machines in a major research hospital'.[5]

THE BRAIN-AWARE PHILOSOPHER

So Descartes was no abstract metaphysician. On the contrary, he delved deeply into the sinews and vessels of the human body that 21st century philosophers at best only theorise about. The major work which he wrote as a result of his anatomical investigations and mechanistic thinking was the *Treatise on Man* (which was published posthumously together with the *Treatise on Light* – containing his cosmology and ideas about the nature of light – as *The World*.) In the *Treatise on Man* Descartes presented an entirely mechanical account of the operation of the human body that encompassed all of its functions, from respiration through limb movement to digestion. But also, and most notably, perception.

The balance of entries in the *Treatise on Man* makes it clear that Descartes expended considerable effort on dissecting brains. He describes their tissue as being 'soft and pliant'[6] and gives accounts of the organ's major structures. His work did not always display the level of accuracy we associate with modern empirical science (this was, after all, the birth of scientific method) and his explanation of the body was permeated with assumptions derived from the medical orthodoxy of the day. For example, he perpetuated a widely-held idea derived from antiquity that the brain's chambers or 'ventricles' were filled with, and even kept inflated by, a quasi-gaseous 'vapour' having psycho-physical properties known as 'animal spirits' despite the evidence which must have been before his eyes that they were in fact occupied by cerebro-spinal fluid.

However, with the level of direct acquaintance with the brain and its major structures that he undoubtedly acquired, it would be fair to say that Descartes became the first neurologically-aware philosopher of the scientific age. Furthermore, unlike most 21st century thinkers Descartes had actually plunged his hands into the grey matter of the brain's tissues. This meant that in assembling his account of perception, Descartes was – perhaps more than anyone had previously – confronting the issues faced by any account that explains perception in terms of brain processes. This was all the more the case because the mechanical form of explanation that he was committed to providing is such a simple one that it leaves no room for vagueness or uncertainty. The consequence was that Descartes was required to dig deep into his resources of creativity in order to find an account that satisfied all of the demands that he was seeking to meet. The resulting theory of perception gave expression to a logical form of explanation which has provided the template for nearly all subsequent brain-based accounts of perception.

NERVES AND SENSORY QUALITIES

The basic problem confronted by all brain-based accounts of perception is this. The body is equipped with a number of peripheral sensory organs and a set of nerves which transport the signals that they generate into the brain. This architecture indicates that perception is an event which occurs in the interior of the brain. But as physical structures nerves cannot convey the sensory qualities of colour, sound, taste, smell and warmth themselves which a perceiver becomes aware of, from the external domain directly into the inner regions of the brain. Thus the quality of colour cannot be transported down the optic nerve and that of sound cannot travel along the auditory nerve, and so on. This means that in order to account for man's undoubted experience of sensory qualities, any brain-based explanation of sensing has logically only a restricted number of options. They are firstly, to identify processes within the brain caused by the action of the nerves as a result of which the sensory qualities might be recreated internally. Or secondly, to identify processes which might operate in such a way as to symbolise or represent the external qualities.

REPRESENTATIONAL THEORIES

Either branch of this decision-tree implies that a pattern of processing arises inside the brain which mirrors that of the qualities apparent in the external world and makes the world perceivable to the subject. The internal pattern represents the arrangement of things in the external world and for this reason theories based on this idea are called 'representational theories'. Descartes' account of how man perceives was just such a representational theory. Indeed, it was arguably the first to provide a detailed articulation of the logic of representationalism from an 'inside–the-skull' perspective.

As with Descartes' cosmology the details of his account of perception bear little relation to modern ideas about the operation of the central nervous system. For example, electric current was little understood as a phenomenon (although static electricity had been known since the Ancient Greeks) so it was impossible for him to conceive that nerves might operate by the transmission of electrical signals.

Instead, Descartes held that the nerves formed a system of tubes that distributed the life-giving vapour of 'animal spirits', which he claimed was manufactured by the pineal gland in the centre of the brain, to the muscles of the body. This vapour was created in such profusion that it filled the principal ventricle of the brain under pressure 'in the same way that a moderate wind can fill the sails of a ship'.[7] An array of pores – essentially valves – on the inner surface of the chamber controlled which nerves the vapour flowed down and Descartes thought that the body's muscles were caused to inflate by the animal spirits thereby producing movement of the limbs.

Moreover, according to this picture of man as an 'earthen machine', nerve tubes connecting the eyes, ears and other sense organs to the brain contained fine fibres linked directly to the pores. This meant that when sense organs were stimulated the resulting disturbance would be transmitted transversely along the fibres – just as, he pointed out, 'when you pull on one end of a cord you cause a bell hanging at the other end to ring at the same time'[8] – and all of the pores connected to the sense organ by the fibres would be caused to open. This meant that the array of stimulation on the sense organ would be replicated as a pattern of pore openings on the inner surface of the brain's principal chamber.

As Descartes described it, in the case of light entering a person's eye from an arrow (the ends and midpoint of which he defined as points 'A', 'B' and 'C'):

'Thus, owing to the different ways in which the rays exert pressure on the points 1, 3, and 5, to trace a figure on the back of the eye corresponding to that of object ABC, as we have already said, it is evident that the different ways in which the tiny tubes 2, 4, 6, and so on are opened by the fibres 12, 34, 56, etc., must also trace it on the inside surface of the brain.'[9]

The pattern of pore openings also gave rise to a matching pattern of outpouring 'animal spirits' on the pineal gland because the vapour exuded outwardly directly towards those pores that were opened. It was, argued Descartes, these patterns on the pineal gland which amounted to man's sensory experience of the external world, encompassing his sensing of 'colours, sounds, smells and other such qualities'.[10] They did so by symbolising or representing the sensory qualities. So, crude though it was in its use of a pneumatic model of mechanism, in the end Descartes had an explanation of sensing which took this outline form:

• Incoming signals such as light or sound waves gave rise to disturbances in the sense organs. The pattern of these disturbances reflected the form and arrangement of objects in the external world.

• A further causal chain of events flowed down the nerves and led to the patterns being replicated in processes within the brain. Because these brain-events maintained the same pattern as those in the sensory organs – and ultimately as that of the external world – they were able to represent to the perceiver features of the external world, including its sensory qualities.

But Descartes was concerned to make clear that representation did not require resemblance. In the *Treatise on Light* he was careful to point out numerous situations in which the 'ideas' or sensations that man has need not resemble the 'objects from which they proceed'[11] (an example being the fact that the feeling of a tickle bears no similarity to the feather which may have been its cause). So his representational theory of human perception did not claim that sensory qualities were replicated within the interior of the brain in order to make the external world apparent to the perceiver. Rather, brain processes – in this case the pattern of out-flowing animal spirits on the pineal gland – could represent sensory qualities (as well, he asserted, as other external features such as edges, size, movement and so forth) without needing to resemble them.

As the foundations of the scientific world-view, based on an atomistic understanding of the nature of matter, were being put together in the 17th century, Descartes had the key insight of working out an explanatory logic for man's perceptual capabilities which took into account the fact that man himself was made of nothing more than atomic matter. Its essence was that brain processes could symbolise or represent events in the external world and thus enabled man to become aware of the world and its contents.

As science has evolved over subsequent centuries and our understanding of the brain as a structure of inter-connected nerve-cells has deepened, so Descartes' representational schema has increasingly provided the template on which theories of perception as a brain-centred process have been built. These representational theories have provided an alternative way of understanding perception to the natural view which carry immense weight, because they are compatible with the scientific world-view, thanks to the logic which Descartes supplied.

4. Newton Against the Natural View

Isaac Newton took the next step in developing an alternative picture of perception to the natural view. He did so by providing an explicit role for the sensory qualities in the brain-based account that had been established by Descartes. This came about as a result of his experimental investigation into the nature of light.

Newton took an interest in studying light, as well as his better known inquiries into the dynamics of moving bodies, from early in his career. According to the husband of Newton's niece, John Conduitt, as a young man 'then not twenty-four' Newton 'bought at Sturbridge fair a prism to try some experiments upon Descartes' book of colours & when he came home he made a hole in his shutter & darkened the room & put his prism between that & the wall'.[1] This account shows that Newton was aware of the work of Descartes from a relatively young age, a fact which is also evident from the notebooks that he kept throughout his studies at the University of Cambridge, where references to Descartes' ideas figure large. One undergraduate at the university recorded that there was 'such a stir about Descartes, some railing at him, and forbidding the reading him as if he had impugned the very Gospel. And yet there was a general inclination, especially of the brisk part of the University, to use him...'[2] From the familiarity which Newton displayed with the French philosopher's ideas one must judge that he was well established as part of the 'brisk' set.

Optics was not a new science. Euclid had described the laws of reflection in Ancient Greece and Descartes himself had been responsible for significant advances in the subject. But in the main it had been confined to understanding the geometry of light rays as they underwent refraction or reflection. To date all ideas about the nature of light itself had remained entirely speculative. Notably this was true of Descartes' own views, which were drawn from his cosmological theory, to the effect that light consisted in small fast-moving particles that had been formed as a residue of the cosmic grinding process by which all elements were originally created. Newton, in contrast, decided to subject light to a battery of experimental procedures with the aim of making it yield up its inner nature. The eventual success of these experiments brought about a quantum leap forward in man's understanding of light and its relation to colour, and an enormous advance in the science of optics. His commitment to the task was illustrated by the fact that some of the procedures were undertaken on his own eyes. On one occasion he stared at the sun for so long that he had to shut himself up in the dark for several days afterwards in order to rid himself of the resulting 'fantasies'. On another he inserted a 'bodkin' between one of his eyeballs and its socket 'as near to the backside of my eye as I could' in order to experience the effects of altering the curvature of his retina.[3]

ANNUS MIRABILIS

But the majority of Newton's experiments on light carried fewer risks than these. Some may have been conducted at Cambridge but in the summer of 1666 the plague swept through England and the university was temporarily closed down. Newton returned to his home village of Woolsthorpe in Lincolnshire where the ensuing isolation saw something of a miracle of intellectual creativity on the young thinker's part, often described as his '*annus mirabilis*'. For it was during this period that the fall of an apple in his garden is thought to have sparked off the train of thought which culminated in his concept of a universal force of gravity. It has been popular at times to cast doubt on this image as one of the 'myths' surrounding Newton but as a past grower of apple trees I suggest that there is a good chance that it may in fact be true. Even in the absence of a breeze, old English varieties of these splendid trees tend to spontaneously drop their fruits during the period of ripeness and when they do the apples, especially larger varieties such as Bramley, smack to the ground with a pronounced '*thwack*' sound. It is a noise which could not have failed but attract even a dreamy young Newton's attention. No mechanical forces are exerted on an apple to make it fall in such a situation, so an inquiring mind such as his might well have asked, why does the apple not simply float in the air next to the tree once it has become separated from its branch? It would be hard to find a more graphic illustration of the pull exerted by the earth on all objects.

SCIENCE OF COLOUR

But it wasn't just gravity that Newton worked on in the rural peace of Woolsthorpe. The exact chronology of his early optical experiments is uncertain[4] but by his own account, 'in the two plague years of 1665-1666...(he) had the Theory of Colours'.[5]

Following Descartes in using the simplest of equipment – prisms and lenses – which he arranged at a 'hole in the Window-shut of a dark Chamber', Newton was eventually able to show two key things in his experiments. Firstly, that colour was not as he put it 'in' light. A popular idea at the time held that colours were modifications of light which occurred during episodes of reflection of refraction.[6] This concept may have been an attempt to account for the spectra of colours that are visible when sunlight is reflected from thin films of oil or is refracted by prisms. Newton demolished it in what he considered an '*experimentum crucis*'. In this, he isolated one of the colours of light in the spectrum created when sunlight was passed through an initial prism and then directed the light ray through a second prism. The result showed that, contrary to the idea that refractions 'add' colour to light, no further chromatic alteration occurred when the already-coloured light passed through the second prism. Another experiment of note also contributed to the same conclusion that colour was not a modification of light. This was one in which Newton shone coloured light (obtained again by isolating the individual spectral colours created by a prism) onto opaque objects in order to see whether its being reflected by them 'added' a further colour to the light. Today, in the 21st century, when coloured light bulbs can be readily purchased at hardware shops and we are used to seeing objects adopting the hue of the light which is shone upon them it is not surprising to us that it did not. But in the mid-17th century sources of light were restricted to the sun and candles, and coloured light was a relatively rare commodity. So Newton was doing something new in even investigating the properties of coloured spectral ('homogeneal' as he called it) light, and the results of his experiment cannot at the time have been readily predictable. With typical thoroughness Newton shone coloured light on: 'Paper, Ashes, red Lead, Orpiment, Indico Bise, Gold, Silver, Copper, Grass, Blue Flowers, Violets, Bubbles of Water tinged with various Colours, Peacock's Feathers, the Tincture of Lignum Nephriticum, and such-like.'[7]

As he reports all of the above objects, 'in red homogeneal Light appeared totally red, in blue Light totally blue, in green Light totally green, and so of other Colours. In the homogeneal Light of any Colour they all appeared totally of that same Colour.'[8]

This experiment was significant in another way too. For it may have raised the question in Newton's mind as to whether colour was an intrinsic property of objects. After all, normally purple violets and blue-and-green peacock's feathers had appeared red under red light, and blue under blue light. So in what sense were they actually purple or blue-and-green? Newton himself never explicitly linked such a question to the experiment in his writings. However, it may be significant that his description of the experiment in his work *The Opticks* was immediately followed by a section in which he declared that colours in objects are nothing but 'a disposition to reflect this or that sort of Rays more copiously than the Rest.'[9]

A further point established by Newton in his experiments was that sunlight was not an homogenous entity as had previously been thought. Instead he showed that it was a mixture of components of light having different refrangibilities (the amount by which rays can be refracted), each component being associated with an individual colour in the spectrum. There was a direct connection between degree of refrangibility (or as we now know, wavelength) and colour. Using only the primitive set-up of a lens and a prism at a hole formed in his window shutters Newton was able to split sunlight into its 'homogeneal' coloured components and then recombine them again to form white light. These are experiments which today form part of every school-child's introduction to the science of optics and demonstrate the extent to which this extraordinary man laid the foundations not only for the physics of gravity but also for our entire modern understanding of light.

Finally, in his experiment to shine coloured light on opaque objects Newton observed that 'every body looks most splendid and luminous in the light of its own colour'.[10] Thus although leeks looked red under red light, and blue under blue light they appeared most bright under green light. Likewise 'Indigo in the violet blue light is most resplendent.'[11] Cinnabar looked brightest under red light, and so on. This enabled him to conclude that opaque objects reflect the various components of sunlight differentially, and that they appear coloured because they reflect more of one component than of the rest.

QUALITY OF COLOUR

However, Newton's experimental findings raised significant questions about colour. The discovery that colour was not a modification of light meant that it had been divorced from the physical signal that brought about the process of vision. If so, what was it that a person was seeing when they experienced colour? In proposing an answer to this, Newton was led to merge his new findings about the composite nature of sunlight with Descartes' representational schema for explaining perception. His modified representational account of colour vision was put forward in a famous 'Definition supplied in *The Opticks*':

'...the Rays to speak properly are not coloured. In them there is nothing else than a certain Power and Disposition to stir up a Sensation of this or that Colour. For as Sound in a Bell or musical String, or other sounding Body, is nothing but a trembling Motion, and in the Air nothing but that Motion propagated from the Object, and in the Sensorium 'tis a Sense of that Motion under the Form of Sound; so Colours in the Object are nothing but a Disposition to reflect this or that sort of Rays more copiously than the rest; in the Rays they are nothing but their Dispositions to propagate this or that Motion into the Sensorium, and in the Sensorium they are Sensations of those Motions under the Forms of Colours.'[12]

The fundamental points here were that:

• Light itself is not coloured (or capable of acquiring colours, as proposed under the doctrine of 'modifications') but only possesses an ability or 'disposition' to cause sensations of colour within man's visual system.

• The colours of objects are 'nothing but' dispositions to reflect certain components of light more than others.

• When this reflected light stimulates man's sensory system it gives rise to sensations of colour.

What Newton had done was to argue that the sensory quality of colour arose internally within man's visual system as a 'sensation'. Sensations are mental entities – they include feelings like pains and tickles – so while his account was a type of representational theory, in that the internal sensations gave the viewer a representation of the external scene, it was distinct from that of Descartes.

On Descartes' account sensory qualities, such as colour, were only represented in the brain by processes which did not resemble them. But according to Newton the colours apparent to man in visual experience were actual mental entities which arose somewhere within his sensory system. We can usefully distinguish between these two approaches to providing a representational account of perception by calling those of a Newtonian type which claim that sensory qualities arise in the mental arena 'quality-based' representational theories, while those of a Cartesian persuasion which make no such claim can be called 'quality-free' representational theories.

CENTRE OF EXPERIENCE

Newton went on to describe in some detail the neurological features of the visual process. He was even prepared to identify the precise location in which the 'picture' which composed the visual experience of an object – presumably made up of colour 'sensations' – arose in the mass of nerve cells that constituted the visual system. This was, he claimed, the point at which the fibres of the optic nerves from each eye crossed-over before proceeding into the brain (known today as the 'optic chiasm'). As he wrote in the *Opticks*:

> 'Are not the Species of Objects seen with both Eyes united where the optick Nerves meet before they come into the Brain, the Fibres on the right side of both Nerves uniting there, and after union going thence into the Brain in the Nerve which is

on the right side of the Head, and the Fibres on the left side of both Nerves uniting in the same place, and after union going into the Brain in the Nerve which is on the left side of the Head, and these two Nerves meeting in the Brain in such a manner that their Fibres make but one entire Species or Picture...'[13]

Furthermore the analysis of sensory experience in terms of 'sensations' applied not only to the visual sense. Thus he explained the case of hearing in the following terms, 'Do not...Vibrations of the Air, according to their several bignesses excite Sensations of several Sounds?'.[14] By implication a similar account might be presumed to hold for the other senses.

THE BISHOP AND THE PRISM

Newton's package of ideas about vision and colour – and sensing more broadly – as presented in the *Opticks* had important consequences. The book was, first of all, highly influential. This was in part due to its ground-breaking approach to the subject matter but also because Newton chose to write it in English, making its impact much wider than if he had written it in the Latin more commonly used at the time for scientific and intellectual books.[15] So Newton's quality-based representational theory, with its concept of sensory qualities as 'mental' entities, made a deep impression on intellectual circles in the English-speaking world – one that persists to this day.

Another related consequence was to give impetus to the idea that sensory qualities might not be a fundamental part of the external world. As we have seen, Newton himself never explicitly drew such a conclusion. However, it was drawn on the basis of his experiment of shining coloured light on opaque objects by the philosopher Bishop George Berkeley (1685-1753).

In 1734 Berkeley published a work called *Three Dialogues Between Hylas and Philonous* which, like Newton's *Opticks,* was designed to be a popular account of his own thought. Amongst a range of questions, the *Three Dialogues* directly addressed the issue of whether colour was an inherent property of objects. Berkeley pointed out that this could not be so for a number of reasons. These included the fact that colours were transient and that many animals in all likelihood saw different colours from humans. But above all else (speaking through the character in his dialogue, Philonous, who broadly-speaking represented the author's position) Berkeley argued:

'Add to these the experiment of a prism, which separating the heterogeneous rays of light, alters the colour of any object; and will cause the whitest to appear of a deep blue or red to the naked eye. And now tell me, whether you are still of opinion, that every body hath its true real colour inhering in it'.[16]

The only possible conclusion was that colour could not be an inherent property of physical objects. Or, as articulated by Hylas, the other participant in the dialogue:

'I own myself entirely satisfied that they are all equally apparent; and that there is no such thing as colour really inhering in external bodies, but that it is altogether in the light.'[17]

So whether or not Newton himself shared this view, a widely derived conclusion from his writings – as a result of the broad accessibility of his own and Berkeley's books – was that the qualities of colour, sound smell and so forth were not only entirely mental entities (as sensations) but also did not exist in the external realm as inherent properties of physical objects.

A BRIEF HISTORY OF THE MENTAL IMAGE
Another consequence of Newton's thought about sensing was the stimulus that it gave to ideas about a mental image. A significant figure of the period was the philosopher John

Locke (1632-1704), who was an acquaintance of Newton's and fellow member of the Royal Society, which provided a forum for the most creative intellects of the time to meet and discuss their ideas. Locke undoubtedly knew of and gave consideration to the implications of Newton's 'new theory of light and colour'. He is best known today for having developed a systematic statement of the difference between the so-called 'primary' and 'secondary' qualities. (Primary qualities being those such as shape, size, motion and number which are inherent in objects whether they are being perceived or not, and secondary qualities being those like colour, sound, taste and smell which are perception-dependent and 'in truth are nothing in the objects themselves, but powers to produce sensations in us.'[18]) But Locke was also one of the first to make explicit the two-dimensional nature of the visual image. In his *Essay Concerning Human Understanding* he wrote:

'When we set before our eyes a round globe, of any uniform colour, v.g. gold, alabaster, or jet, 'tis certain, that the idea thereby imprinted in our mind, is of a flat circle variously shadowed.'[19]

The rest of his writings leave room for doubt as to whether Locke believed in anything as sophisticated as the modern notion of a mental image (the matter is still debated by Lockean scholars). Nonetheless, this passage would have introduced many of his readers to the concept of visual experience as dependent on an internal two-dimensional visual projection of the external scene – a notion which Locke proceeded to reinforce by comparing the 'idea imprinted in our mind' to a painting.[20] So in a few simple words Locke conveyed to the post-classical world the concept that the visual component of the mental image might be 'flat'.

PLASTIQUE BERTRAND

Locke's idea evolved over subsequent centuries into the modern concept of the mental image. It may be no coincidence that the next step in this process took place in the early 20th century, when Western man's visual environment was undergoing a period of transition as man-made materials became more commonplace[21] and colour was starting to be more frequently observed on the perfectly 'flat' surfaces offered by moulded materials such as plastics. In this context, the philosopher Bertrand Russell was able to persuade his readership – without significant dissension – that the two-dimensionality of Locke's coloured mental image could encompass all sensory qualities. This occurred in his 1914 paper 'On the Relation of Sense-data to Physics', during the course of which Russell set out a picture of sensory experience as taking the form of 'patches of colour, sounds, tastes, smells, etc., with certain spatio-temporal relations.'[22] A 'patch' is a two-dimensional entity, so in one simple move Locke's vision of the mental image as made up of a flat array of colour had been expanded to incorporate 'patches' of sound, taste, smell and in principle all of the other qualities.

MOVIE IN THE BRAIN

Russell's conception of the mental image as a two-dimensional mosaic composed of patches of all of the sensory qualities lives on today, but with the addition of a dynamic element, to allow for the fact that objects are usually perceived in motion. This gives rise to a 'movie' metaphor, as we see in the following passage from neurologist Antonio Damasio (in which he addresses the problem of how there can be a sensory quality-laden mental representation in the brain):

'Solving this problem encompasses, of necessity, addressing the philosophical issue of qualia. Qualia are the simple sensory qualities to be found in the blueness of the sky or the tone of sound produced by a cello, and the fundamental components of the images in the movie metaphor are thus made of qualia. I believe these qualities will be eventually explained neurobiologically although at the moment the neurobiological account is incomplete and there is an explanatory gap.' [23]

CONTRADICTIONS OF COLOUR

The progression from Descartes through Newton, Berkeley and Locke to the modern day has shown how representational theories that claim sensory-qualities occur within the mind offer a potent alternative to the natural view. In direct contradiction to the natural view, this type of theory as it has come down to the modern world carries the implication that sensory qualities are not inherently real features of the physical world. If they are located anywhere according to such theories it is in the mind of the perceiver in the form of sensations making up a quality-rich mental image.

The effect of this position has been the more powerful because it emerged in history at the same time as (and was thus fully consistent with) the foundational concepts of science. Its origins lay in the post-classical rediscovery of atomism, and its development later continued with Newton's investigation of optics which laid the basis for our modern understanding of light. Inevitably, as both neuroscience and man's conception of the physical world have developed over recent centuries, this picture – in stark contrast to the natural view – has been able to link in well with developing scientific ideas. Also it is entirely free of the conflicts with the atomistic picture of matter from which, as we saw in Chapter 2, the natural view suffers. For example, it is untroubled by the argument that atoms and molecules do not have the properties of colour, smell, sound and so forth which the physical world appears to possess. For it holds that such sensory qualities are not genuinely in the external world, but rather are sensations within the perceiver's mental image.

So for all of these reasons, the Newton-inspired 'quality-based' representational explanation of perception has provided a powerful alternative to the natural view. As a result, in the early 21st century we live at a time when the contradictions in perceptual understanding are greater than ever. With the continuing growth in science, the potency of this type of representational account has never been stronger. Yet man's inclination to live by the natural view continues to show that it too has lost none of its appeal.

PART II:
FROM COLOUR TO
CONSCIOUSNESS

5. Mega-mind and Micro-mind

We have seen that ideas about the sensory qualities divide into two camps. On one hand are concepts, derived from the work of Newton, Berkeley and Locke, to the effect that colour, sound smell, taste and so forth are solely mental entities generated in the mind of the perceiver by the action of stimuli such as light and sound on the sense organs. On the other hand are those which make up the outlook which I have called the 'natural view'. These assert that sensory qualities are real external phenomena which exist in the physical world whether or not it is being perceived. The issue which divides these two 'camps' could hardly be more fundamental; it is the very existence of sensory qualities in the physical realm. This highlights the fact that while by definition they are theories of sensing, each of the two ways of conceiving of the sensory qualities has implications for our understanding of reality. For the Newton-inspired picture claims that physical reality does not contain sensory qualities whereas the natural view is adamant that it is replete with them.

Argument Without End?

On the face of it, it may seem that an issue as basic to man's understanding of reality as this stands little chance of being resolved. After all, philosophers have been struggling with the 'great questions' for thousands of years and few, if any, definitive answers have emerged. So what hope can there be of settling an issue as fundamental as whether or not sensory qualities exist in the physical world? Especially given the great simplicity of colour, sound, smell and taste, which makes it virtually impossible to get a conceptual 'handhold' on any matters concerning them.

Perhaps it should come as no surprise, therefore, to discover that philosophical debate over this question has existed for many centuries, and not just in the Western tradition that stems from Ancient Greece. Thus in the philosophically advanced civilisation of classical India around 500–1000 AD (half a millennium before the birth of Newton) we find an ongoing argument involving numerous schools of thought which centred on the status of the sensory qualities,[1] although it was not phrased in those terms. The sophistication of thought in the period is evident from the various explanations for visual illusion put forward by three of the rival schools, each having distinct implications regarding the status of colour.

For example, adopting a Newton-like position, the Yogacara school of Buddhism argued that when a person mistakenly sees a white shell as a piece of silver the appearance of silver must be a mental entity which 'shares the character of such "internal" episodes as pain or pleasure.'[2] The view of the Advaita Vedanta school, on the other hand, was that the appearance of silver in such an illusion must belong to a 'third realm' of objects which were 'neither existent nor non-existent'[3] – in effect neither physical nor mental, although such categories were not used at the time. Finally, perhaps the most subtle analysis of all was due to the realist schools of the Prabhakara Mimamsaka and the Nyaya. They suggested that such an illusion did not require an explanation in terms of how two colour appearances – white and silver – could be sensorily present at one moment to the perceiver. Rather, they asserted that white existed on the shell and was experienced directly as such by the perceiver. Silver came into the illusion, they argued, through a perceptual judgement made by the perceiver that the shell was silver, but not through any actual sensory experience of silver-ness. The judgement was instead produced by the intervention in the visual act of a non-conscious or 'concealed remembering'[4] of previously-seen silver objects – the latter being stimulated in the perceiver's mind by the similarities between the colours silver and white.

Given, then, the ancient and seemingly endless nature of the debate over whether sensory qualities are objective or subjective, should we conclude that its solution is forever unobtainable? Is man condemned to eternal speculation about the location and status of sensory qualities?

Fortunately a little lateral thinking helps us see that this need not be the case. A fissure arises in the otherwise insurmountable metaphysical fortifications surrounding the issue, because each of the two accounts of the sensory qualities implies its own concept of mind. By comparing the relative merits and defects of these concepts we obtain a means of evaluating the depictions of the sensory qualities themselves. We can deduce, as it were, that which is obscured by the metaphysical wall surrounding the topic. We can, in effect, prise open the fissure a little and peer through it with restricted vision, at least far enough to determine which of the two pictures of the sensory qualities may be right. For, as it turns out, one of the concepts of mind has considerably more to recommend it than the other.

MEGA-MIND
Consider first the picture of the sensory qualities which is derived from Newton and Berkeley. This makes essentially two claims:

a) That sensory qualities are sensations generated in the perceiver's mind.

b) That the qualities of colour, sound, smell, warmth, taste
and so on, at least as we normally understand the terms, do
not exist in the physical world.

The mind, on this picture, must act as the crucible within
which all of the colour, sound, smell, taste and so forth which
a person experiences is created. At the same time the external
physical realm, being devoid of colour and other qualities, remains
a blank emptiness of atoms and space. Also there is no suggestion
on this picture that there exist in the universe any other types
of device which have the same capability as 'minds' to express
sensory qualities. In other words, in the entire cosmos the mind
alone (which Newton, at least, seemed to equate with the brain –
in terms of its sensing capacities) is deemed capable of performing
this act of filling in the empty shapes and forms of the world with
hue and tonality. So this account of the sensory qualities grants an
absolute level of uniqueness to minds. It also deems them to be
devices of extraordinary sophistication. For within the bounds of
a person's 'mind' all of the richness of the universe must somehow
be conjured. In effect, the entire weight of explaining sensory
qualities is shifted onto the 'mind' under such a scheme. As such,
this apparently unique type of device is required to perform a
feat which must render all who consider it speechless. For the
generation of sensory qualities is comparable to the creation
of the physical universe in the pantheon of ultimately creative
events. This is a concept of the mind, then, of the grandest order.
It is the idea of a hugely sophisticated device which is at once
cosmically unique and in possession of capabilities which are of
the most fundamental nature. We might appropriately call it the
'mega-mind' theory.

MICRO-MIND
The contrast with the concept of mind suggested by the natural
view could hardly be starker. According to this account of the

sensory qualities, colours exist on the surfaces of objects (or, in the case of translucent colours, through their volumes). Sounds, smells and so forth are distributed throughout the physical realm. In other words, the sensory qualities are located exactly where they appear to be in acts of perception. What is more they exist in these positions whether or not they are being observed. Thus the surface of a lemon is yellow both when it is being examined by a prospective purchaser in a grocer's shop and early in the morning when the shop is empty of customers.

According to this way of understanding the sensory qualities, the mind is required to contribute little if anything to the process of perception. If sensory qualities are already 'out there' prior to and during every act of perception, then in explaining perception there is no need to call on the mind to generate them as internal 'sensations'. Indeed to do so would be redundant, for all of the subjective sensory elements that are encountered in perception (the colours, sounds, smells, tastes, warmth, touch and so on) pre-exist in the external physical realm. It is almost as if the mind has to contribute nothing to the process in order for a person to experience the vividness contained in the world. So, the markedly different and much less sophisticated concept of mind which follows from this picture is one which is stripped down to the barest minimum of contents. This is a 'micro-mind', as opposed to the massive structure of a 'mega-mind'. Indeed it becomes possible to conceive here that the mind is a device of the greatest order of simplicity, which perhaps barely even exists at all in the conventional sense.

6. Mind and Body

Having now looked at the concepts of mind that are suggested by the two principal accounts of the sensory qualities, the next step will be to evaluate their impact on the mind–body problem. This is the notoriously challenging puzzle of how the mind and body are related. More specifically, the problem concerns questions about how bodily events can give rise to mental ones, for example, how the stubbing of one's toe causes a feeling of pain. It also concerns questions about the way in which mental events can give rise to bodily ones, for example, how one's decision to move an arm makes this actually happen. Ultimately, there is the question of how the complex processing carried out by the billions of neurons in a person's brain gives rise to an entire mind. Or, as philosopher David Chalmers has recently put it: 'Why doesn't all this information-processing go on "in the dark", free of any inner feel?'[1]

THE QUEST OF DESCARTES

Since the birth of the neurosciences in the 18th century, with the discovery by Italian physiologist Luigi Galvani that nerves operate by means of electricity, man has made huge strides in understanding the principles underlying the operation of the central nervous system. But these advances have only highlighted how little insight has been gained into the relation of the mind to the body, and the mind-body problem has become one of the key intellectual challenges of our times. Indeed some philosophers – the so-called 'mysterians' – now argue that the problem can never be answered. They claim that man's cognitive faculties are inadequate for him to understand the nature of his own consciousness.

The mind-body problem was originally articulated, at least in its broadly modern form, by Descartes, as a by-product of his search (famously inspired by a dream during a sojourn in a stove-heated room) for sound philosophical foundations for the nascent field of science. The method of doubt which he was subsequently led to invent brought him to the idea that the only thing that man can be certain of is his own existence as a thinking being. This Descartes encapsulated in the well-known phrase: 'I think, therefore I am' (*Cogito ergo sum*).

From here the distinction between the mind and the body followed almost immediately. For on the one hand, as Descartes said, a person is 'a thinking and unextended thing'; while on the other their body is an extended and unthinking thing.[2] Descartes was thus able to conclude that: 'This I (that is to say, my soul by which I am what I am), is entirely and absolutely distinct from my body, and can exist without it.'[3]

THE HARD PROBLEM OF CONSCIOUSNESS
In the original Cartesian rendering of the mind-body problem the mind is conceived of as a thinking substance (*res cogitans*). Although, having said that, Descartes' conception of 'thinking' was a broad one, including as it did, doubting, affirming, denying, knowing, loving, hating, willing, desiring, imagining and perceiving.[4]

In more recent times, the mind-body problem has been rearticulated in terms of consciousness. This has involved a shift of focus to the specific problems presented by mind as the centre of subjective, primarily perceptual, experience. This is evident in one of the best known modern discussions in the field: David

Chalmers' contemporary restatement of the problem as the 'hard problem of consciousness'. According to Chalmers, the issues raised by consciousness divide into a number of 'easy' ones and one very substantial 'hard' one. The 'easy' issues are those which present relatively less of a challenge because they are amenable to scientific investigation, and include the need to provide explanations for:

- The ability to discriminate and react to environmental stimuli.

- The ability to integrate sensory information.

- The ability to focus one's attention.

- The ability to carry out deliberate actions.

- The difference between being awake and asleep.[5]

The 'hard' problem, in contrast, differs from these by an order of magnitude in the level of challenge that it presents. Its subject matter concerns, quite simply, the 'problem of experience.'[6] As he eloquently puts it:

'When we think and perceive, there is a whir of information-processing, but there is also a subjective aspect…there is *something it is like* to be a conscious organism. This subjective aspect is experience. When we see, for example, we *experience* visual sensations: the felt quality of redness, the experience of dark and light, the quality of depth in a visual field. Other experiences go along with perception in different modalities: the sound of a clarinet, the smell of mothballs. Then there are bodily sensations, from pains to orgasms; mental images that are conjured up internally; the felt quality of emotion, and the experience of a stream of conscious thought. What unites all of these states is that there is something it is like to be in them. All of them are states of experience.'[7]

The problem of experience, the 'hard problem of consciousness', is:

'...the question of how physical processes in the brain give rise to subjective experience. This puzzle involves the inner aspect of thought and perception: the way things feel for the subject. When we see, for example, we experience visual sensations, such as that of vivid blue. Or think of the ineffable sound of a distant oboe, the agony of an intense pain, the sparkle of happiness or the meditative quality of a moment lost in thought. All are part of what I am calling consciousness. It is these phenomena that pose the real mystery of the mind.'[8]

The reason that the problem of experience is so hard, argues Chalmers, is that experience cannot be explained solely in terms of neurological functions in the way that the subject matter of the 'easy' problems can. Thus when a person sees the colour red, the experience is accompanied by various events in his or her visual system. For example, light of a certain wavelength will have struck his or her retina and this will have activated a cascade of neurological functions to discriminate and categorise the resulting information. But none of these facts by themselves explain the subjective 'inner feel' of a vivid sensation like redness as it is experienced by the perceiver. 'Why doesn't all of this information processing go on unassociated with any "inner feel"?' asks Chalmers. There is, he points out, an 'explanatory gap' between the subjective experience on the one hand, and on the other, any explanation in terms of neurological functioning. What is more, there seems in principle to be no way in which this gap can be bridged. Thus for any attempted account of consciousness in terms of neurological functions – no matter what level of detail is given of the brain's internal processes – the question 'And why is it accompanied by the experience of redness, or saltiness (or whatever)?' will always remain.

How Hard a Problem?

How do the two views of the sensory qualities and their consequent concepts of the mind bear on the hard problem of consciousness? There can be a certain ambiguity inherent in the way in which the hard problem is presented. For in one sense it may be understood as the question of how the felt qualities of experiences such as redness or the smell of mothballs *themselves* arise inside the network of neurons which comprise the brain. In another, it may be understood as that of how our *awareness* of such felt qualities can arise in the brain, without prejudging the issue of where the actual qualities themselves are located.

On the first of these interpretations – which Chalmers himself does little to dispel by talking of conscious experience in terms of a rich 'inner life', the 'inner feel' of perceptual experience and mental images that are 'conjured up internally'[9] – the sensory aspects of the mind-body problem reduce to the problem of the relation of sensory qualities to the physical matter of the brain. While on the other, the possibility is held open that sensory qualities may be located externally to the brain, and the hard problem becomes a question of accounting for how the 'whirring' of the internal neurological mechanisms enable man to gain an awareness of them.

The significance of these interpretations lies in the fact that they correspond almost exactly to the consequences for the hard problem of the two concepts of mind that we have been considering.

Logical Dead End

Take first of all the 'mega-mind' concept suggested by the type of representational theory that is derived from Newton. Recall that this proposes that colours and other sensory qualities arise within the arena of the mind and enable us to see a version of the world apparently endowed with qualities. Newton's own account and all representational theories since Descartes have been brain-based, claiming that the mental arena is instantiated in the brain. So in so far as it is derived from such accounts the mega-mind concept is linked to the notion that sensory qualities occur somewhere (in a manner unknown) *within* the neuronal fabric of the brain.

But such a scheme encounters two fundamental types of difficulty. Firstly, unlike matter, sensory qualities have no objectively measurable quantitative features. So, there is a stark difference between the matter of which the brain is made and the sensory qualities which on this picture are supposed to arise within it. On the one hand, the neurons and even the molecules of which they are composed have properties such as mass and electric charge that can be measured objectively with scientific instruments. The same, however, cannot be said of, say, the colour red or the smell of mothballs.

While it may be possible to make subjective assessments of the relative intensity of the sensory qualities, they do not have quantitative properties of the same sort as the mass or electric charge of atoms and molecules. This means that it would be impossible for there ever to be a 'bottom-up' explanation of how sensory qualities arise in the matter of the brain. In science such accounts typically explain one high level quantitative feature in terms of a lower level one, as when the chemical reactivity of a molecule is accounted for in terms of the behaviour of the electrons in its constituent atoms. But in this case, there are no high level quantitative properties to explain in terms of the mass, electric charge, chemical reactivity and so forth of the physical components of the brain such as its cells and the electric signals flowing between them.

So the non-quantitative nature of sensory qualities means that the mega concept of mind – together with the representational theory from which it is derived – is logically-speaking a dead-end when it comes to the hard problem of consciousness. It would in principle be impossible ever to provide the explanation of the occurrence of sensory qualities within the matter of the brain which this concept of mind requires.

CONTINUOUS QUALITIES VS. GRAINS OF MATTER

There is a further difficulty for the type of representational theory that is associated with a concept of mega-mind. This was identified by the philosopher Wilfrid Sellars and has come to be known as the 'grain problem'.

The grain problem points to the discrepancy between the continuous nature of sensory qualities as appearances and the 'granularity' of any structure within the brain onto which they might conceivably be mapped. Thus colours are typically experienced as continuous and homogenous fields of visual quality on the surface of objects. Likewise sounds, such as the pure tone of a cello or a synthesizer, may be experienced as unbroken streams of auditory quality. But this aspect of sensory qualities is at odds with the discrete nature of the neurons and synaptic junctions – and at the smallest scale of atoms and molecules – that make up the functional units of the brain.[10]

The grain problem highlights another fundamental difference in nature between sensory qualities and the brain. The former are indivisible and continuous, whereas the latter is made up of vast numbers of component units. Once more this indicates the impossibility of achieving an account of how sensory qualities could arise within the matter of the brain.

MEGA HARD PROBLEM

It emerges that there can be no hope of solving the hard problem under the mega-mind concept. True, man's understanding of the role of the billions of neurons which constitute the human brain is as yet far from complete. But there is no prospect that any future development in neuroscience will span the conceptual barriers presented by these two problems. It is conceivable that at some stage in the future an understanding of matter as an entirely geometric structure of space-time may be developed (this was the vision pursued by Einstein in his later years and also by American physicist John Wheeler[11]). A potential consequence of such an account might be that the grain problem would hold less weight because matter – including the matter of the brain – would be understood to enjoy a similar form of continuity as sensory qualities. But no conceivable conceptual development offers the prospect of getting around the mismatch between the qualitative nature of sensory qualities and the quantitative nature of matter.

Under the mega-mind concept and its associated type of representational theory, the hard problem of consciousness represents therefore not so much an 'explanatory gap' as an explanatory *gulf*. Effectively, the problem becomes as hard as it could ever be. It becomes insoluble.

MICRO HARD PROBLEM

The contrast with the implications of the micro-mind conception could not be greater. Recall that here all colours, sounds, smells and so on are understood as occurring in the physical world, where they do so even when unperceived. In an act of perception the mind is called on to do virtually nothing to enable these external sensory qualities to be perceived. To appreciate the impact of this on the hard problem it is useful to revert for a moment to John Locke's language of primary and secondary qualities. For what this account of the sensory qualities asserts in effect is that the status of colours, sounds, smells and so forth (the 'secondary' qualities in Locke's parlance) is the same as that of his 'primary' qualities of shape, size, number and so on. Under the natural view both are thought to inhere 'in' objects when not perceived. But we would not say, and Chalmers did not argue, that there was a 'hard problem' about our awareness of the primary qualities of shape, size and number of objects, which are intrinsic to the external realm and as such part of the 'outer' domain. This is one of the 'easy' problems having to do with awareness and categorization.

Likewise Locke took it as a given that with an understanding of the basic mechanics of perception in place, no special problem was presented by man's awareness of the primary qualities. In the same way, under the micro-mind conception that the sensory qualities are objectively real there would be no 'hard problem' concerning our consciousness of these qualities. According to this picture, sensory qualities such as colour, sound, smell and taste pre-exist our perceptions of them in the external world and therefore the only problems in accounting for our sensing of them would be 'easy' ones having to do with how we are able to attend to, memorise and categorise such qualities.

The micro-mind concept, and the natural view with which it is associated, makes the claim that what Chalmers regards as 'inner feels' are in fact 'outer feels' in the external world. With that key difference the hardness of the hard problem simply evaporates. It would be replaced solely by a set of 'easy' problems concerning the brain's ability to attend to such external features.

Philosopher Nigel J.T. Thomas has summarised this effect succinctly (using the term 'qualia' for sensory qualities):

> 'If the color realist is right, and colors are "really" out there on the surfaces of objects, rather than in here in our minds, then we no longer need to postulate color qualia in the mind or brain, and the problem of explaining how brain states could embody such qualia disappears.'[12]

ULTIMATE LOGIC

These dramatically different consequences for the hard problem allow conclusions to be drawn concerning the merits of the respective accounts of sensory qualities from which the 'mega-mind' and 'micro-mind' concepts are derived. We have seen that the mega-mind conception (linked to the Newton-inspired approach to sensory qualities) produces a gulf between the mind and the body that is so large as to be unbridgeable. Nigel J.T. Thomas is also alert to this point, saying: 'After all, if sensory qualities, qualia, are really in minds rather than in the physical world, then there must be minds, distinct from the physical world, in order to accommodate them.'[13]

In contrast, the micro-mind notion associated with the natural view's account of sensory qualities eliminates this gap. If it leaves any problems concerning our awareness of the sensory qualities they are only of the 'easy' kind.

This tells us that in terms of ability to yield a coherent explanation of reality that does not require two irreconcilable domains (of consciousness and matter), the natural view's account of sensory qualities performs better than the quality-based representational approach. The conclusion must be that the natural view's account of sensory qualities is a superior one to the quality-based type of representational theory that is derived from Newton.

PART III:
COLOURS AS
RELATIONS

7. Rescuing Direct Realism

All other things being equal, the natural view provides a superior account of the sensory qualities than the quality-based representational theory, which takes its inspiration from Newton. The former reduces the problems associated with explaining man's access to the sensory qualities to 'easy' ones, while the latter magnifies the gulf between mind and body to an impassable chasm. So why aren't philosophers falling over themselves in the rush to praise the natural view? In fact, if anything, the opposite is the case. The theory which is modelled on the natural view, known as 'direct realism' (or sometimes 'naïve realism') is probably the least popular position amongst philosophers of perception. Why? Simply because all other things are not equal. The natural view suffers from a host of problems that quality-based representational theories of perception manage to evade. We have already encountered some of the problems faced by the natural view, particularly its inability to integrate with the scientific picture of physical matter, but there are many more.

PROBLEMS FOR THE NATURAL VIEW
Consider, for example, the fact that objects don't always appear the way that they are in reality. Thus if you press gently on the side of one of your eyeballs, this has the effect of seeming to displace all of the objects in view to one side (this is a milder

and considerably safer version of Newton's experiment to apply pressure at the back of his eye with a 'bodkin') yet in reality the objects have not moved. A difference has been made to arise between what you see and the arrangement of things in the external world. Or, more prosaically, if you look through rippled glass of the type often found in bathroom windows, the outside world looks contorted and bent, whereas in fact it is unmodified.

The mismatch between physical reality and the way that it appears in such cases is easy to explain under theories of vision such as Newton's, where colour is held to be a mental artefact. Here the discrepancy can be attributed to colour sensations having been produced in a location within the observer's visual representation which does not correspond to their true position in external reality – due to the abnormally distorted path of the incoming light rays. But according to the natural view the colour that we see is actually located externally on the surfaces of objects. It is hard to reconcile this with the difference between the known location of those physical surfaces and their apparent one in such cases.

Then there is the fact that perception can give rise to illusions. As, for example, when a person who has taken a powerful drug such as LSD suffers extreme visual hallucinations. But there are also visual illusions of a less spectacular nature that are experienced by us all, and the best known of these is perhaps the after-image.

After-images occur when you have stared for too long at a bright source of light, such as a lamp, and your field of vision acquires an image of the light-source having the appearance of a 'blob' of yellow which hangs in the air wherever you look. Contrary to the natural view, an 'after-image' of yellow, which appears to 'float' in the middle of your field of vision, is certainly not a case of seeing a real surface colour currently on a real physical object. Surely the easiest way to account for it is, as Newton would have done, simply a sensation of yellow? A further difficulty for the natural view is that from the point of view of sensory experience there are no apparent differences between the colours which make up illusions and those on physical objects themselves, so it can be argued that if the former are sensations then all experienced colours may be – including those which appear to be located on the surface of objects.

To compound the problems faced by the natural view there is also the fact that the colour of an object may not always appear the same to all observers. This leads to situations where a single object can be seen as having multiple colours. Take, for example, a red cherry on a cocktail stick which is within sight of a number of people as they gather round a table of drinks at a party. Anyone who has normal vision (and is in a normal state of mind) will see it as red. But there may be some amongst the party-goers who are red-green colour-blind and therefore see a different visual quality on the cherry in place of red. Others again may see it as the colour pink because, having taken an hallucinogenic drug, they hallucinate the cocktail fruit to be a piece of glowing pink plastic atop the stick. Others still may have decided to examine the cherry through the blue-tinted glasses in which their drinks have been provided, and so see it as purple in colour. There may be some who view the cherry from an angle at which it catches the light from a wall-mounted spot-lamp that the host has fitted with a yellow light bulb in order to lend 'atmosphere' to the party. As a result they see it with an orange colour. Such a multiplicity of colour appearances on a single object is not consistent with the simple 'one colour per object-surface' idea that is associated with the natural view.

MULTIPLE APPEARANCES AND SELECTIVE PERCEPTION

We saw earlier that the natural view lacks the internal coherency of a fully developed theory. Yet its primeval appeal has nonetheless encouraged thinkers to distil its essential features into more systematic accounts of perception. This has led to the creation of a number of theories which attempt to defend the natural view's core idea, that sensory qualities exist objectively in the external world. I will highlight here two of significance to the thrust of this work. The first is due to an American philosopher of the early 20th century, E. B. Holt, who proposed that objects have multiple appearances and that man's sensory apparatus selects from amongst them during acts of perception.

According to this 'selective' theory of perception the natural view is correct in asserting that physical objects are coloured and that it is their surfaces which we experience in vision. (Thus the quality of red exists on tomatoes and is accessed directly by perceivers.) However, it is wrong to say that each object only has a single appearance. In fact, according to Holt, each object manifests *multiple* appearances to the world and a person's perceptual apparatus is able (without any conscious awareness on the perceiver's part) to select a cross-section from amongst them during perception. If the perception is illusory then, as a result of causal events in the perceiver's eyes and brain (for example, red-green colour-blindness), a certain cross-section of appearances will be selected. Whereas if it is veridical (that is, physically-accurate) this merely means that a cross-section has been selected which happens to correspond with the physical arrangement of the object in the world. But in all cases, including those of illusions, the appearance which is experienced by the perceiver is located on the object and is not a mere phantom in his or her head.

In this way Holt's theory provides an explanation of perception which is in tune with (at least the spirit, if not with the letter of) the natural view, and which can account for such otherwise baffling phenomena as illusions. He demonstrates that with the application of a certain amount of lateral thinking, ideas can be generated that breathe philosophical life into the natural view.

Thus the account of the sensory qualities which in practice defines most people's perceptual lives does not have to represent the philosophical dead-end that it might otherwise seem. Also Holt's ideas highlight the potential of the idea of a multiplicity of appearances to act as an effective tool for solving the difficulties faced by the natural view. This is a clue which we will have occasion to make use of later in the present book. Unfortunately, this idea represents also the point of greatest weakness in Holt's position. For despite their efficacy, the multiplicity of appearances on which his selective theory relies produces a metaphysically-bloated conception of reality. It is one which is unlikely ever to find acceptance unless independent arguments were found which demonstrated that objects do indeed possess more than one appearance.

REFLECTANCE COLOURS

The other theory stemming from the natural view which I will highlight is a more recent one and concerned exclusively with the sensory quality of colour. It was put forward in 2003 in a highly regarded article in the journal *Behavioural and Brain Sciences* by Alex Byrne and David R. Hilbert entitled 'Color Realism and Color Science'.[1] In the article Byrne and Hilbert endeavour to defend 'the minority view that physical objects (for instance, tomatoes, radishes and rubies) are coloured, and that colours are physical properties.' They also take into account as far as possible recent findings from the science of colour vision. Their conclusion is that it is a 'credible hypothesis' that an object's colour can be identified with the proportion of incident light which it is disposed to reflect at each wavelength in the visible spectrum. This property is known as 'surface spectral reflectance' and so Byrne and Hilbert call their idea 'reflectance physicalism'. By expanding the definition of colour to cater for all light leaving an object not just by reflectance but also through transmission and emission, they show that reflectance physicalism can readily be extended to cover the colours not only of opaque surfaces but also of translucent bodies and coloured light sources (such as traffic lights).

One of the main strengths of reflectance physicalism is that it caters for the puzzling fact that objects retain a constant colour appearance under modest variations of illumination (although as Newton showed, they do not stay constant under major variations). To use Byrne and Hilbert's example, if you bring a tomato indoors from a sunny vegetable patch to the kitchen it does not appear to change colour as the illumination changes from sunlight to the artificial lighting inside. This is because 'the proportion of incident light which the tomato's surface is disposed to reflect at each wavelength in the visible spectrum' does not change during the process. While the amount that the tomato actually reflects at each wavelength in the visible spectrum alters when it is brought indoors under the different form of illumination, the proportion that it is disposed to reflect stays constant, because this is a physical characteristic dependent only on the micro-cellular make up of the fruit's skin. Spectral reflectance information about objects is carried by light which has been reflected from them, and a further strength of reflectance physicalism, according to Byrne and Hilbert, is that empirical work has shown how the human visual system might recover such information from the responses of the photoreceptors in the retinae of our eyes.

In many respects, then, Byrne and Hilbert's reflectance physicalism represents a powerful statement of how colour may simply be a physical property of the surfaces of objects (or their volumes in the case of translucent objects, such as rubies). However, on the negative side, such a conception of colour sees it as intrinsic to those surfaces and volumes and thus exhaustively contained within the (restricted) space that they offer. As we will see in the next chapter, there are grounds for thinking that such a conception of colour may be inadequate.

Colour – Constant or Variable?

All theories of perception modelled on the natural view share the idea that sensory qualities such as colour and sound exist objectively in the physical realm. They hold that colours exist on the surfaces of physical objects and sounds and smells are distributed throughout the physical environment. As mentioned earlier, most theories of this type have come to be called 'direct realism' or in some cases 'naïve realism'.

Many thinkers use these two terms interchangeably, but I will draw a distinction between them here. The theory of 'naïve realism', I will say, refers to the most primitive ideas about the sensory qualities within the natural view. Specifically in the case of colour, I will say that it refers to that component of the natural view which holds that the colour quality of objects is fixed *under all lighting conditions*. It is, then, the idea that limes are green, lemons are yellow and strawberries are red under all colours of light and even in the dark. Thus it is according to the naïve realist component of the natural view that baked beans sealed away from light in their can remain orange. In contrast, the somewhat more sophisticated 'direct realism' represents an enhancement of the natural view's ideas about colour according to which the colour quality of objects genuinely *varies* under different lighting conditions. Direct realism holds that baked beans become temporarily black in their can due to the lack of light and that a lemon becomes green when a blue light is cast on it, and so on.

Of these two versions of colour realism the naïve realist version of the natural view holds little prospect of success. For it rests on the idea that colours are, as it were, 'glued' onto the surfaces of their respective objects (or are embedded throughout their volumes in the case of translucent colours). They thus remain unchanged no matter how the lighting conditions are varied. Such a notion carries a strong sense of artificiality. It also, of course, runs directly counter to man's visual experience. Newton's experiment of shining different coloured light on objects and observing that they changed their appearances showed this. But we see it also in everyday life. For example, when the light of a sunset tinges a landscape with orange hues, and the blue jeans of a person crossing lines of stationary traffic at night are made to look purple in the red light cast by the brake-lights of vehicles. Indeed, the idea that colour is an invariant quality of objects is in such marked contrast with human experience, that a rejection of the 'naïve' form of colour realism seems almost inevitable.

DIRECT ACCESS

On the other hand, the 'direct realist' distillation from the natural view, under which colour is said to genuinely vary according to the hue of ambient light, is fully consistent with experience and can be said to represent the viable element of the natural view's approach to colour.[2]

There is, however, more to the theory of direct realism than simply the fact that colour varies under different lighting conditions. For example, direct realism also adopts the natural view's approach to *all* of the sensory qualities. In other words, in addition to light-dependent colour, the qualities of sound, smell, taste, heat and cold are all held to be real characteristics of the external world which exist within the physical realm whether or not they are being perceived. But direct realism as the term is normally understood, and as I shall use it here, also makes explicit a feature which within the natural view is only tacit. This is that perception gives the subject immediate (hence 'direct') access to such external sensory qualities. This contrasts in particular with the type of representational theory which claims that in perception man only has immediate access to sensory qualities as mental entities (sensations) that represent external scenes to him *as if* they were full of colour, sound and smell.

What is Direct Realism?

Direct realism, as I shall understand the term, can therefore be summarised as follows:

It is first of all the idea that in perception no sensations or other mental versions of sensory qualities intervene between the subject and his or her awareness of external physical objects being perceived. That is not to exclude the possibility that *some* form of representation of those objects may be involved in the process of their being perceived (and may even be necessary for perceptual awareness and instantiated somewhere in the perceiver's brain). It is, however, to exclude the possibility that such a representation is quality-laden.

For the centre-piece of the direct realist viewpoint, as with the natural view, is that the sensory qualities which we experience in our perception of physical objects are real external phenomena and are directly perceived as such. Sensory qualities exist in their locations in the physical world without any dependence on perceivers. They are objectively real features of the physical realm and do not 'blink off' as soon as the perceiver ceases to sense them (by looking away or closing off his or her ears or other sensory organs). The appearance of the physical world is therefore identical whether you are engaged in perceiving it or not. When you perceive (see, hear, smell, etc.) the external world you simply become directly aware of that pre-existing appearance.

Perceptual Pariah

However appealing such a form of direct realism might be, as we have seen it is bedevilled by a series of critical problems. There is also the awkward fact that it fails to integrate with the scientific understanding of physical matter as atomic in nature, and is rivalled by the apparently successful alternative that quality-based representational theories present. In retrospect, it is hardly surprising that direct realism should have become almost the pariah of perceptual theories amongst contemporary thinkers. Yet, as we have seen, it is only this account of the sensory qualities which offers the prospect of a 'micro' concept of mind and thus the possibility of a breakthrough in the hard problem of consciousness. How can such benefits be accrued? Only, it would seem, if a version of direct realism could be developed which overcomes the difficulties which this viewpoint faces. But philosophers have been thinking about the natural view – and hence in one way or another, direct realism – for hundreds of years. As we saw in Chapter 5, the history of such thought goes back at least as far as the Nyaya school of classical India.

So what hope can there be of developing a new form of direct realism containing features which have not previously been considered? Is it conceivable that there is any aspect of sensing which might have been overlooked in all of those centuries of intellectual endeavour? If there is, it surely must lie in the subtly complex nature of the sensory qualities which was revealed in our opening meditations on the beach. For there we learnt that certain colours at least – those of the sky and the sun – are not simply the flat expanses of visual quality that the conventional conception of colour would have them be. Instead they radiate outwardly from their physical locations with an inherent directionality. Likewise sound and smell share the same characteristic of outward manifestation from their physical locales, an idea which also goes against conventional notions that if sound and smell reside at all in the physical realm they do so as simple, self-contained regions of sensory quality.

NEW DIRECT REALISM
The powerful contrast between these distinctive features possessed by certain sensory qualities, and the more conventional conceptions which direct realists have worked with until now,

suggests that there may be mileage in creating a new version of direct realism that incorporates such features. What is more, there are indications that these features may be just the ones that could enable the direct realist viewpoint to overcome some of the key difficulties which it faces.

For example, if the colour of objects in general 'manifests outwardly', this could potentially provide a basis for a direct realist explanation of how multiple observers can see the same object as having different colours – as in our earlier example of a cocktail cherry being observed by many viewers at a party. Thus it could be that objects present outwardly a somewhat different colour to each perceiver, the hue presented to each depending on the conditions of observation (whether the observer is looking through coloured glass, has colour-blindness, and so on). If so, then the many colours apparent on the cherry would not arise separately in each observer's head but instead would be located on the surface of the fruit and would manifest outwardly from there to the individual observers.

By building in the features of sensory qualities that we identified on the beach there may, after all, be a way forward for direct realist accounts.

Certainly the signs are sufficiently positive to warrant a deeper investigation. That being so, it would be good to establish a program of action to examine further the potential of such ideas to improve direct realism. First of all we should look more deeply into the surprising and subtle features of the sensory qualities that were encountered on the beach. We need to know, for example, how widespread the characteristic of 'outward manifestation' is amongst the sensory qualities and more precisely what its nature is. Once a deeper understanding of the sensory qualities has been achieved which takes outward manifestation into account then we can apply whatever findings are established to the development of an enhanced form of direct realism. Then it will be necessary to test the performance of any new version of direct realism it proves possible to build with the conventional array of arguments used against the position. Will it or will it not fare any better?

8. Relational Colours: A New Entity

The first step towards enriching direct realism is to examine how widespread the characteristic of 'outward manifestation' is amongst colours. So far the blueness of the sky and yellowness of the sun have been identified as displaying this characteristic but there are other examples of colour in the natural environment which display it too. One regularly seen by us all is the blackness of outer space. Just as with the colour of the sky, this does not appear to be a visual quality laid out on a surface, as if the celestial blackness were a coating of colour on a vast wall, deep in the nether regions of the cosmos. Rather, in a similar fashion to that of the sky's blue colour, it gives an appearance both of depth and of 'manifesting outwardly' from the remoteness of the universe.

CELESTIAL SPHERE

But can we be certain that the properties of the appearance of the colour of space are not a cognitive artefact produced by our modern knowledge of the vast scale of the universe? After all, in the Middle Ages it was commonly held that the stars were fixed points on the outermost of a set of vast, concentric, crystalline 'celestial spheres' which rotated in space around the earth, carrying the planets and stars with them. Some thinkers of the period may have held that the appearance of the stars, as twinkling points of light, derived from the fact that they were holes punched in the outermost sphere through which shone a 'divine' light that was thought to exist in a realm beyond. If so, they may have also believed that the inner surface of the outer sphere was opaque black, for that is the colour apparent between the stars. But this suggests that they considered the blackness of the night sky to have been a coating intrinsic to the inward-facing surface of the huge, distant sphere – in direct contradiction to how we experience the colour today.

As a matter of general rule, though, it seems unlikely that the difference between a colour appearing to be on a surface or not, and manifesting outwardly or not, can be determined solely by the observer's conceptual framework. Take the sun for example. The intensity with which its yellowness appears to radiate outwardly is so great that it is hard to imagine any system of concepts that an observer might hold altering this aspect of the way in which its colour is perceived. The blackness of space is perhaps a more borderline case, and space as 'the heavens' is the inevitable target for some of man's grandest conceptual schemes. But even here it is possible to find experiences of such intensity (unavailable in the Middle Ages) as to suggest how improbable it is that our modern awareness of the outward manifestation of the blackness of the night sky is a mere cognitive artefact.

Such experiences come from the testimony of men who have been closer to space than most of us are ever likely to get. Consider the reactions of the astronauts William Pogue (Skylab 4) and Charles Duke Jr. (Apollo 16) upon seeing its naked depths close-up during space-walks[1]:

'I had given up trying to use a small flashlight to continue our work in the dark. I raised the visor on my helmet cover and looked out to try to identify constellations. As I looked out into space, I was overwhelmed by the darkness. I felt the flesh crawl on my back and the hair raise on my neck. I was reminded of a passage in the Bible that speaks of the "horror of great darkness".

Ed and I pondered the view in silence for a few moments, and then we both made comments totally inadequate to describe the profound effect the scene had made on both of us. "Boy! That's what I call dark.'" William Pogue

'It was a texture. I felt like I could reach out and touch it. It was so intense. The blackness was so intense.' Charles Duke Jr.

During these walks Pogue and Duke's experience of the blackness surrounding them was clearly raw and overpowering; too much so to be the product of a mere conceptual understanding of the extent of space. In this direct and intense experience of the colour of space Charles Duke Jr. in particular alludes to its possessing an element of manifesting outwardly when he describes it as having a 'texture' and mentions being able to 'reach out and touch it'.

THE SUB-AQUATIC SKY

Another example of a natural colour which manifests outwardly is provided by the underwater environment. Consider the scene which can greet a diver beneath the waves. It occasionally happens in a dive, particularly during a slow descent or ascent, that a diver gets the wonderful opportunity simply to float with neutral buoyancy in mid-ocean. What he or she sees then is that in the distance the transparency of sea-water gradually fades to an opaque blue-green colour. This blue-greenness has both depth and is 'outward facing', in a similar fashion to the blueness of the sky. As in the previous cases it has none of the appearance of a colour on a surface, as if there was a flimsy curtain coated in blue-greenness hanging in the water at a distance. Instead, this is a colour appearance which *manifests outwardly* from the distant water, just as the blue of the sky does from the heights of the atmosphere. The colour of distant water provides a constant visual feature in the underwater world during daylight in much the same way that the sky's colour does in the air. In fact the parallels between the two are so strong that one is tempted even to call the distant blue-greenness visible underwater a 'marine sky'.

Outward manifestation, it seems, is not a characteristic peculiar to the colours of the sky and the sun. It is widespread amongst the colours of nature and, while the marine sky requires specialised equipment in order for it to be observed, the other examples cited here are everyday colours frequently seen in normal life.

A thought prompted by the case of the marine sky, however, is that it suggests that had man been an aquatic creature, he might have considered the characteristic of outward manifestation to have been virtually universal amongst colours. For unlike its aerial cousin the marine sky is almost ubiquitous underwater – it occurs in virtually every direction a diver cares to look and forms the backdrop to most visual scenes. This means that the majority of colour seen underwater has the distinctive characteristic of outward manifestation, in contrast to the terrestrial environment where the majority of colours appear 'flat' and surface-like. An aquatic species of human would have been likely to consider the case of colours on surfaces a peculiarity and that of outwardly manifesting ones to be the norm.

The Shape of the Sky

As it happens, the marine perspective on colour may be the correct one. For there are grounds for thinking that all colours, including surface ones, and not just those of bodies such as the sky and the sun, manifest outwardly. These stem from the fact that there is one thing common to all of the examples of colour which clearly manifest outwardly. That is that they are seen in conditions which offer no visual cues. When we observe solid, opaque physical objects we become aware of their shape by means of various 'cues' which the human visual system discriminates in the wealth of information presented to it in the light reflected from those objects. In particular, we make use of visual elements such as boundaries and differences in colour or tone to distinguish an object from its background and so acquire an awareness of its form. But the sky offers no such cues that could be used to establish its 'shape'. At its limits the sky has no edges, because it has no definable boundaries in the way that a solid object such as a house or an apple does. Nor are there any sharp differences in colour between the sky and any background in front of which it is seen, because it is not seen against such a background. The same goes for space and the marine sky. In each of these cases there may be certain foreground objects – clouds, fronds of seaweed, planets and stars – which establish the distance and hence depth of the coloured entity being viewed. But the entity itself, the sky, outer space or distant seawater is seen without the benefit of visual cues and so remains visually without form. The sun also is seen without visual cues, although for a different reason. Here the sheer intensity of its light – so great that we can stand no more than to glance at it – obliterates all visual indicators regarding the spherical nature of the solar body.

THE TRUE NATURE OF COLOUR

Such conditions are significant for the perception of colour because under them there can be no influence on our experience of colour from an awareness of shape. They, uniquely, provide man with an exposure to colour which is unaffected by any imposition of spatial form. It follows, therefore, that in such conditions man gains his most accurate experience of the sensory quality of colour with regards to its spatial nature. For it is only under these conditions that the spatial form of his experience is not manipulated by the powerful effect of visual cues. And what spatial nature do they reveal colour to have? As we have discovered, they consistently reveal it to have one in which it manifests outwardly.

The true spatial nature of the sensory quality of colour, revealed on those occasions when there is no potentially distorting effect from visual cues, is therefore that it manifests outwardly. If this is the genuine nature of colour in respect of its spatial form, it follows that it must in fact be the spatial nature of colour in all situations. So, despite any appearance to the contrary, even colours in more conventional settings – such as on objects' surfaces – must in reality present outwardly. The red of a tomato's skin, the yellow of a child's bucket, the green of a toy car and all similar colours must not, as they may appear, be confined to their ostensible physical surfaces. If it is not always obvious that colours present outwardly from such surfaces that is because their presentation is obscured for us at a deep level in our visual systems by the visual cues offered by the objects themselves.

PRESENCE ACROSS SPACE

So I argue that outward manifestation is an ubiquitous feature of the sensory quality of colour. But of what, more precisely, does it consist? How can it be explained? Consider next this situation:

It is a balmy summer afternoon and you are strolling through a mature, well-tended garden. There are flowerbeds on either side of the path, densely packed with flowering plants of all kinds. You notice a rose bush freshly in bloom, and you hunch down beside it to get a closer look. You peer in towards a particularly succulent scarlet flower. The fine downy texture of its intimate flesh, the delicate structures of stamens and petal roots are extraordinary. But most sensational of all is the vibrancy of colour displayed by the dusky petals. Their rich, vivid redness is almost overpowering in its depth, unmatchable by any man-made colour. The redness seems to radiate out towards you, across the space between the petals and your eyes. You begin to wonder...

How is it that you are able to be aware of the redness? Of course, light from the sun is reflected from the petals into your eyes, which triggers a cascade of neuronal activity in your brain. But the more significant point is, how can you, or indeed any person, be aware of a colour quality that – according to direct realism – occupies a location that is distant from the point at which you are? The colour quality is in one place, you are in another, and yet despite the physical separation an awareness arises in the place you occupy of a colour quality that is some distance away – how?

In most scientific thinking about the physical world its elements are conceived of in highly compartmentalised terms. Each car, house, brick, planet and atom is thought of as a self-enclosed, discrete entity, occupying its own space and independent of all surrounding objects. Each is fundamentally a self-contained item, which does not depend on any others for its existence. If an object's neighbours ceased to exist there is nothing that demands that it must do so also.

This understanding in terms of compartmentalised objects, which science obliges us to use for the physical realm, encompasses the scenario of a rose and its observer. Separated in space, they must be conceived of as discrete entities with no fundamental interdependencies holding between them. When it comes to understanding the colours involved in the scenario, therefore, within the terms of such a scheme it is intelligible that the redness of the rose's petals be 'on' them, wholly contained within their physical boundaries. It is also intelligible that if there are any colours associated with the person carrying out the observation they be confined within the space of his or her body. (These could be the colours of his or her skin, eyes and hair, for example, or perhaps – at least for the sake of argument – some colours in a visual image within his or her brain.) But the scientific conception of objects does not have a mechanism to allow for the sort of presence that the red of a rose's petals, or any other coloured surface, has across the gap of space between the two.

BOX WORLD AND COLOUR UNCONFINED

The access that a person's eyes have to remote colour in this and most other cases of colour vision goes against the whole grain of discreteness in physical thinking. It represents a connection between two localities in the universe of a kind that is difficult to assimilate into a world which supposedly only contains entities as self-contained 'boxes'. The colour, or at least its presence or the observer's awareness of it, is not spatially limited in that way. Here we can see what the phenomenon of outward manifestation of colour truly consists of. Whether it concerns surface colours, as in the case of the rose, or the colour of entities such as the sky, the fact that colours manifest outwardly involves them in *presenting across distances of space*.

In the case of the blueness of the sky, the colour quality presents from above the clouds across a gap of space equivalent to the thickness of the atmosphere down towards the earth. In the case of the scarlet rose's petals, the redness presents across a gap of space from the surface of the petals to at least as far distant as locations occupied by possible observers around it. In cases where the outward manifestation of a colour is obvious because it is not swamped out by visual cues, what we are really saying, then, is that man has a heightened awareness of this feature that the colour presents across space. With the colours of the sun and the sky, in particular, this heightened awareness is especially clear. The evidence that this is so is apparent in the visual certainty that we have that such colours cannot be on surfaces.

EXHAUSTIVE CONTAINMENT

The currently prevailing way of conceiving of external surface colour qualities conforms to the compartmentalised mode of thinking characteristic of man's physical conceptions of the world. That is to say, such colours are normally thought of as fully contained within the (severely limited) space offered by those physical surfaces. The outer red colour quality of a tomato, for example, is commonly considered to be fully located within the thin region of space offered by the skin of the fruit and not to have any element which reaches out beyond that into the surrounding space. For future reference I will call this view the idea that colours (and, as we will see in future chapters, also other sensory qualities) are 'exhaustively contained' within their supporting object.

Yet it is precisely such a failure to be exhaustively contained and instead a 'reaching out' beyond the confines of a colour's physical location which occurs in presentation across space. A good example of a conception of colour as exhaustively contained within its supporting objects is provided by Byrne and Hilbert, whom we looked at in the previous chapter. According to them, colour is a physical feature (the proportion of incident light which objects are disposed to reflect at each wavelength in the visible spectrum) that is entirely intrinsic to the surfaces and volumes of physical objects. But this means that their theory has difficulty accounting for man's awareness of colours at a distance. According to their thinking, colours are entirely bundled up in objects' surfaces and volumes, yet the viewer, as we have just noted, is invariably some distance from these. Reflectance physicalism provides no explanation for such remote awareness of intrinsic surface colour qualities across gaps of space.

BLOCK OF SPACE

Accounts that conceive of colour as exhaustively contained in its supporting object offer a picture of this sensory quality parallel to that of sensory appearances such as weight and hardness. These features cannot be sensed at a distance from objects and therefore give the impression that they are inherent to objects. So, reverting to the example of the proverbial can of baked beans, when observed from a distance by a perceiver this provides no sensory access to its hardness and weight. If the sensory quality of colour was exhaustively contained in objects then the same would be true of it. Thus the blueness of a baked bean can's label would be confined to the region of the label in the same way that the weight of the can is intrinsic to the matter of the can itself, and the colour would be obscured to any viewer by the distance of space separating him or her from its location (in the same way that the distance makes it impossible to sense the can's weight). This latter point demonstrates two things. Firstly that no direct realism of the sensory qualities based on the simple notion that the quality of colour is exhaustively contained in objects could ever account for the fact of remote visual experience. Secondly, that in coming to a full understanding of the nature of presence across space we will almost certainly need to consider the connection between sensory qualities such as colour and space itself.

PRESENCE ACROSS OUTER SPACE

It is worth noting two things about the phenomenon of presence across space. Firstly, we have seen that although outward manifestation is only obvious to us in cases such as the colour of the sky and the sun, it in fact occurs in all instances of colour. Also it is clear that remote awareness of colours, such as in the example of the scarlet rose, is characteristic of virtually all human colour visual experience. This means that the phenomenon of presence across space is something which must characterise all colours evident in the physical world, both those on surfaces and those which more obviously manifest outwardly such as the blueness of the sky. Secondly, presence across space can in some cases occur across very wide expanses of empty space. The sun, for example, is 149,600,000 km away and its colour quality clearly has no difficulty in presenting across that distance. Colour from the stars presents from even further.

Next, let us ask ourselves the question, 'To which of the array of points in its surroundings may the colour of a typical object be presented?' Take the scarlet rose from earlier, which was bathed in sunlight. How far into the surrounding space may the redness of its petals be presented, and are there any points which do not have the colour presented to them? Concerning distance of presentation, we can deduce from the examples of the sun and stars that there are no practical limitations of distance for colour presentation within the observable cosmos. It seems that if light encoded with information about an object can travel to a point, no matter how distant, then the object's colour can be presented there. On this basis, the red of the rose may be presented as far into the surrounding space as light from it can travel without encountering an obstacle (although cosmic distances at which the light loses its energy and thus information content may perhaps provide an outer boundary). Equally we know from visual experience that when the rose is observed from any point in the surrounding space which has a clear path of view to the flower, its redness may be seen. This suggests that there are no positions in the volume of space around the plant to which its redness is not presented, except where light travelling away from its surface is impeded by opaque objects in the vicinity. This is consistent with the general idea that colour is presented to wherever light from the object can travel.

A New Paradigm – Relational Colours

The idea that colour presents across gaps of open space suggests a very different viewpoint from the box-like conception of exhaustive containment typified by ideas such as the reflectance physicalism of Byrne and Hilbert. The latter, conforming to the compartmentalised scheme of thought characteristic of science, conceives of colour as a discrete entity fully contained within the boundaries of its supporting object. Whereas the idea that colour presents across space sees it as a *connecting event* between the location from which the presentation occurs and the points to which the presenting is done. In short, and here we arrive at a key conclusion, presentation across space carries with it the notion that *colours are relationships*. But these are not abstract relationships of the type often discussed by modern philosophers such as 'being to the left of' or 'having twice the weight of' and so forth. On the contrary the fact that a colour presents across space is a claim about what occurs in the space of the physical world. They are, then, very much concrete relationships within the physical realm.

103

Nor is the presentation of colour a matter of subjectivity or in any way dependent on observers. (Note that throughout my description of the phenomenon I have said nothing about any requirement for observers to be present.) The relationships which we must deem colours to be are entirely physical relationships. That is, they are relationships which span actual physical extents of space.

ATOMS, MOLECULES... RELATIONSHIPS

This idea means exactly what it says. I am suggesting that the relationships which we call 'colours' are a previously unrecognised type of entity within the physical universe. According to this line of thought such space-spanning relationships must join the pantheon of entities recognised as constituents of the universe alongside atoms, molecules and other particles and forces.

INTER-RELATIONAL COLOURS

The general idea that colours are 'relationships' has been proposed before by a number of thinkers.[2] But in fact this is only in the very different sense that colours might be inter-relationships between objects and viewers arising in a relation of co-dependency between the physical characteristics of external objects and the internal states of the viewers.[3] This sort of view is often seen as providing a half-way house between the full-blown objectivism of direct realism and the subjectivity of positions such as Newton's. The history of such inter-relational accounts of colour goes back as far as Aristotle. In his *De Anima* (III. 2) he argued that sensing in general worked through a 'merging' of objects' potentialities to be coloured, flavoursome, odorous or whatever with the potential possessed by our senses to respond to such features. In this meeting of compatible potentialities were realised the actualities externally of, colour, sound, odour and so on, and internally of the experiences of seeing, hearing and smelling.

More recently, the case has been made by Evan Thompson that

a wide range of findings in cognitive science are consistent with his own version of an inter-relational account of colour. This is based on a so-called 'ecological' understanding of human and animal sensory psychology (of the type pioneered by J.J. Gibson).

According to Thompson, as a property of surfaces on and around which animals and humans conduct their lives, being coloured 'corresponds to surface spectral reflectance as visually perceived by the animals.'[4] Equally, as a property of the ambient light in the air (or water) through which humans and animals move, it corresponds to 'the spectral characteristics of the illumination as visually perceived by the animals.'[5] Finally, Edward Averill has put forward arguments in an attempt to show that colour terms can only be specified in the context of the population of 'normal observers' who see the colours together with the environment which provides the viewing conditions under which they are 'normally' seen.[6]

While the inter-relational approach to explaining colour has some appeal because of its apparent fusion of subjectivism and objectivism, it suffers from a significant problem. It relies on the assertion that the sensory quality of colour exists in the external physical environment (necessary for it to fulfil any claim to objectivism about colour), or alternatively it says little of significance about the qualitative content of man's visual experience. But in the former case, no mechanism has ever been put forward by its proponents to account for the alleged dependency of external colour qualities on man's neurologically-based visual system.

It also fails to offer an account for the distinctive feature of sensory qualities that has been identified in the present chapter – that they present across space.

9. Sense Qualities as Relations

So far we have established that far from being confined to its location (as suggested by conventional object-based conceptions) colour presents across spans of space. From this we derived the notion that colours are in fact relations in the physical realm, holding between the locations from which the presentation occurs and those to which the presenting is done. In this chapter we will move on to examine the other sensory qualities. In our opening meditations on the beach it appeared that they too showed signs of outward manifestation. Could it be that they also are relationships?

TURN ON THE HEAT

Let us start with heat. Consider firstly the type of radiant heat given off by the glowing coals of a fire. This form of heat quality is simultaneously felt as being located within the body of the coals and also as 'radiating' or presenting outwardly from there. It is apparent that it is not fully contained within the boundaries of its supporting object (here, the glowing coals). It does not conform to the paradigm of exhaustive containment, under which the quality of heat would be locked inside the space of the physical object with which it is associated. Quite clearly, the quality of heat is not bound within the limits of the pieces of coal. Rather it manifests outwardly from there, in much the same way as the colour red does from the petals of roses and blue does from the sky. We also feel radiant heat from the sun on our bodies and this illustrates the fact that with heat such presence (as with colour) can span across very substantial distances of empty space.

Heat energy can also be transmitted from objects in another form, by conduction. However, even then, the essential characteristics of its associated sensory quality remain the same. A good illustration of this occurs when you take a bath and the heat from the water is conducted directly into your body. Once immersed in the hot water, its enveloping warmth is clearly experienced as directed. It has a direction which is presented neither up into the air, nor out towards the walls of the bathtub but rather, in towards your body. In other words, once again the heat quality is not exhaustively contained in its supporting object, in this case the bath-water. Rather it is presented in the direction of energy conduction outwards from the volume of water towards your body. We can conclude that the sensory quality of heat (whether the result of radiant or conductive transmission of energy) is not exhaustively contained in its associated object but presents outwardly from it. In all of these respects it behaves exactly as was found to be the case with colour.

OBJECTIVITY OF ODOUR

What about smell and taste? The paradigm of colour as exhaustively contained within the surface or volume of its supporting object, would tend to suggest a similar 'bound into matter' model for the sensory qualities of taste and smell. But as soon as one attempts to think about scent and taste as being objectively located within the physical realm in such a locked-in form, a host of profoundly puzzling questions ensue. For example, is the tart taste of a lime actually embedded in the flesh of the fruit? Does a sweet possess its sugary sweetness before it has been unwrapped and licked? Is the aroma of a freshly baked croissant in the bread itself? Or alternatively is the croissant's distinctive smell in a little cloud of molecules around the doughy crescent?

From the outset one senses that there is something seriously wrong at the conceptual level in the thoughts which all of these ideas express. The difficulty that we have in conceptualizing the location of the sensory qualities of smell and taste gives a strong indication that these sensory qualities are not fully intrinsic to their apparent locations, as the paradigm of exhaustive containment would have it. But the same conclusion is also arrived at when considering how we actually experience taste and smell.

For example, consider what happens when you smell an espresso coffee. It appears as if the aroma of the beverage extends from the little white mug and all around it towards you. The experience is not one in which the aroma seems to be intrinsically embedded inside the liquid contents of the mug, nor even confined to a 'cloudy' region of air above it. The smell is experienced as if it suffuses outwardly from its source location. In other words, employing the terminology that I have made use of previously, it is *presented* from the coffee to the surroundings and is detected as such by the perceiver. Certainly, smell is a less tangible sensory quality than, say, colour. Yet it seems clear that it is by no means tightly located in or around its source object. Instead, it presents outwardly across space from there in much the same way that colour does. The distances over which aromas may be presented cannot, clearly, be anything like the cosmic ones which colours in some cases are. Instead, just as the range of colour presentation is determined by light which carries colour-related information, that of smell must be determined by the diffusion distances through air which odour-conveying molecules are able to travel.

THE LONELINESS OF THE LONG-DISTANCE FLAVOUR

Taste is a sense which has much in common with smell, as both are dependent on the detection of diffused chemicals, and it is even possible that certain tastes may be qualitatively identical to certain smells.[1] As a first suggestion therefore it seems plausible to propose that a similar picture to that which has been sketched out here for the sensory quality of smell could be built up for taste. Just as the sensory quality of smell presents outwardly from sources of olfactory molecules, so the quality of taste may do so from food such as nuts and sweets, as sources of gustatory molecules. But in the latter case this would only be when these items were being eaten and the flavour molecules had had sufficient time to mingle with the saliva present in the mouth of the person consuming the portion of food. The flavour quality would then be able to be presented through the medium of the digestive fluids in the mouth, in conjunction with the diffusion of gustatory molecules in much the same way that odours are presented through the medium of air in conjunction with the diffusion of olfactory molecules.

However, there seem no possible grounds on which to consider that a completely dry and unmasticated piece of food might have a taste. Thus it appears inconceivable that a nut in its hardened case or a sweet still unopened in its wrapper should have an intrinsic flavour. (Juicy fruits such as oranges and raspberries might seem better candidates for intrinsic possession of the sensory quality of taste because they do not suffer the inherent drawback of a lack of moistness. But in the final analysis possession of intrinsic flavour appears equally impossible even in such cases. This is because of the difficulty of saying where within the physical structure of such fruits a supposedly inherent quality might reside.)

As regards the sensory quality of taste, the only reasonable position, therefore, seems to be that items of food acquire their quality of taste only when mingled with saliva in the mouth of a person consuming them and possibly also being subject to a degree of breaking down through mastication. The item's taste quality then manifests outwardly from it through the digestive juices in the same way that smell does from its source objects and colour does from the sky. But an interesting secondary question then arises. Is the actual presence of a conscious perceiver tasting the food required? Could items of food acquire tastes simply by being immersed in bowls of saliva and perhaps also being chopped up in order to encourage their surface molecules to diffuse through the liquid? (Such an image may seem a little unsavoury but the question is pertinent.)

As with the other sensory qualities, the answer in part depends on the broader philosophical position that is adopted regarding the objectivity of sensory qualities in general. But there seems to be nothing intrinsic to taste as a sensory quality which singles it out as uniquely unable to exist in the absence of a perceiver. So, the answer would seem to be that quite possibly yes, an edible item, such as, say, a sweet, might acquire a taste simply by being placed in a bowl of saliva. There may be a delay following its initial submersion, lasting until the digestive enzymes in the saliva have released molecules from the sweet's surface and allowed them to diffuse within the liquid. However, once that has happened the taste of the sweet would be conveyed through the saliva in exactly the same way as odour is through the air – albeit in a liquid rather than a gaseous medium. And just as smell presents outwardly from a source object, so it would seem plausible to conceive here that taste does as well. (This is consistent with our experience of flavours, which rarely appear bound up and confined within the items being consumed, but instead suffuse throughout the mouth during the process of eating in the manner of aromas in the air.)

That being the case, we can conclude that within the confines of its liquid medium, whether or not a perceiver is present, the sensory quality of taste presents outwardly just as colour does – and all the other qualities so far examined.

Bubble of Sound

Next let us think about sound. After colours, sounds are probably the most common of all of the sensory qualities that we experience. Indeed, in this noisy modern world it can be difficult to think in general terms about the nature of sound as a quality. To overcome this hindrance, consider an example of a sound which is isolated from the incessant auditory background.

Take the case of a cannon being discharged on the summit of a nearby hill. The 'bang' that results is not experienced as a discrete, self-contained 'bubble' of sound that pops up around the cannon remotely from the listener on the hilltop. On the contrary, the sound quality reaches out from the summit towards surrounding locations. Sound presents itself across the intervening gap of space between the hilltop and the surroundings in precisely the same manner as the redness of a rose reaches out from the surface of its petals across the intervening space to the viewer. Once acknowledged for an isolated incident such as this, it becomes clear that the characteristic of presenting across space is a feature of all sounds. Whether it be a dog barking at the far side of a valley, the noise of a pin dropping nearby on a hard floor or the cumulative sound of an orchestra filling an auditorium, sounds never occur as 'bubbles' of noise exhaustively contained within the immediate region of space around their source events. Instead, like colours, they span the gaps of space (in some cases quite large) to the surroundings.

LAYER OF FEEL

Finally there is touch. The natural view's concept of touch differs from its notion of the other sensory qualities because it implies that there is no externally existent sensory quality for this sense. Thus, take the tactile property of hardness. Solid objects are invariably hard but it is not part of our everyday picture of them to consider that they have a quality of 'hardness-feel' which coats their external surfaces invisibly and which is revealed at man's finger-tips when he touches them. Likewise objects may be textured in the sense of being rough or furry, but they are not thought to be coated in 'roughness-feel' or 'furriness-feel'.

There is no fully systematised conception within the natural view regarding the feelings that man becomes aware of through his sense of touch (in large part because, as we have seen, the natural view does not represent a coherent and systematic body of theory). However, if there were one, given that such feelings are not held to be external 'coatings' of 'touch quality', it would have to be that they arise as internal bodily sensations. Thus the feelings induced by the texture of a surface such as an animal's fur coat could be thought of as thousands of minuscule internal bodily sensations, analogous to 'tickles', generated in one's fingertips as each of the hairs of the coat microscopically 'tap' against one's skin when one's finger is drawn across it.

In fact, regarding touch, this type of position seems to be the only one available. For in the case of texture it would be absurd to suggest that invisible layers of 'furriness-feel' or 'roughness-feel' or any other grade of 'feel' existed on the outer surfaces of textured objects, and that these were what gave rise to our awareness of texture. In the case of smooth surfaces (at room-temperature) the same is true, but in an even more obvious way. For in touching un-textured objects all that we feel at our fingertips is a sort of emptiness. This is, I suggest, a tactile form of transparency: a positive tactile experience with no textural content (analogous to visual transparency which is a positive visual experience with no colour content). The tactile transparency that results from touching smooth objects makes it inconceivable in such cases that they should be coated in layers of 'emptiness-feel'. We have already observed that under the natural view it is not considered to be the case that solid objects are coated in any 'hardness-feel', and this seems to be the only reasonable position to take regarding the felt quality of 'hardness'. So all of the possible 'touch-feel'-types – furriness, roughness, hardness and the emptiness offered by smooth surfaces – have been ruled out from existing externally as objective surface qualities on objects which man is made aware of through his sense of touch. All of the feelings associated with the sense of touch must therefore be internal bodily sensations similar in nature to itches, tickles and pains.

TOUCHING EXISTENCE

The lack of an external sensory quality gives touch something of a special status. Because no sensory qualities intervene between the perceiver and object this sense offers man a particularly direct and complete experience of objects. While the same is true of vision and visual transparency, in the case of touch this sense of directness is further accentuated by the physical contact that man has with the object, so leading to the experience that in touching things he is directly exposed to their physical existence. It may have been a sense of this direct exposure to existence, as acquired through tactile contact, which led Samuel Johnson to use the example of kicking a boulder (a characteristically over-dramatic and no doubt painful variant of 'touching') as the basis of his 'refutation' of Berkeley's idealism concerning matter. This is how his companion James Boswell recorded the event in his *Life of Johnson* (1791):

> 'After we came out of the church, we stood talking for some time together of Bishop Berkeley's ingenious sophistry to prove the non-existence of matter, and that everything in the universe is merely ideal. I observed that though we are satisfied his doctrine is not true, it is impossible to refute it. I never shall forget the alacrity with which Johnson answered, striking his foot with mighty force against a large stone, til he rebounded from it, "I refute it thus."'

TOUCHING YOU

There are many ways in which the sense of touch is unique, being active where the others are (largely) passive, and also possibly – it is plausible to suggest – having been the earliest of life's senses to have evolved.[2] This sense even gives the lie to the widely expressed view that sense experience is necessarily private. Or, at least, it shows that idea up as resulting from a bias in favour of quality-based representational theories. This can be seen by touching together two of your fingers. How many 'feelings' of touch do you become aware of? Only one. What, therefore, should you expect to happen when you and someone else touch fingers (or any other body parts) together? Surely that only one such feeling occurs, and that it is shared between you. In other words, that the feeling is public and not private.

We only believe such feelings to be private, because of what is known about the operation of the nerves, and because of the quality-based representational theory which has been grafted onto this knowledge. Thus it is known that at the moment of contact between the fingers of separate individuals, different nerve impulses are generated in each person's body, and that they travel separately up into each individual's brain where – according to Newton-inspired doctrines of sensing – distinct *sensations* of touch are produced. In other words, the alleged privacy of sensing derives more from current representational models of how sensory appearances arise than from anything logically or structurally inherent in the act of sensing or sensory qualities themselves.

SENSE QUALITIES AS RELATIONS

Having surveyed all of the sensory qualities we can now draw the conclusion that it is not only colour which presents across space. In fact all of the sensory qualities that we have looked at show evidence to one degree or another of this phenomenon. The only human sense which stands out as in any way different is touch. But here the difference lies not in the fact that the sensory quality does not present across space. Rather it is that touch is not associated with an external sensory quality at all (and hence the issue concerning presence across space does not arise).

This in turn means that it is true to say of *all* sensory qualities (remembering that the sense of touch does not involve access to one) that they are connecting events between the location from which the presentation occurs and the points to which the presenting is done. Hence all sensory qualities are relationships. Thus when the quality of sound, for example, presents from a cannon discharged on a hill-top to the surrounding countryside it forms a connecting event – and thus a physical relationship – from the cannon to those surroundings. Likewise the aroma which a croissant presents to the space around it forms a connecting event, and hence relationship, between it and the points of space in the neighbourhood. The taste of a sweet that is being eaten forms a connecting event, and relationship, to all of the points in the consumer's mouth to which molecules from the confectionery diffuse.

The sensory qualities occur, and thus the corresponding relationships exist, in the physical realm whether or not any conscious perceivers are in the vicinity. (Even in the case of taste the quality arises, as a relationship, when items of food are immersed in digestive fluids alone.) More generally, we can say of all of the sensory qualities that they are relationships of presenting embedded within the fabric of the physical world. The only qualification that needs to be made to this is that the sense of touch does not involve access to a sensory quality and so does not reveal such a relationship in the way that the other senses do.

A Small Adjustment to Physics

All of the sensory qualities without exception, then, are physical relationships which span regions of space. As such they must be considered (as we noted for colour in the previous chapter) to be entities within the physical universe alongside fundamental particles, fields and forces. But surely this is a thesis of physics? Is it not a theory which could only be accepted following extensive experimentation, not to mention work on its implications for the mathematical theories that underpin physics? Furthermore, if found to be true would not the addition of new fundamental physical entities such as these have consequences that would ripple widely through the whole framework of our scientific beliefs?

Fortunately such implications for science do not follow from acknowledging the existence of relational sensory qualities as new forms of entity within the physical realm. The sensory qualities, as appearances, have none of the objectively measurable characteristics typical of objects of scientific study such as atoms, rocks and planets. Thus, for example, the colour orange has no electric charge, the taste of sweetness possesses no gravitational attraction and the smell of freshly-mown grass has no mass. In general, the sensory qualities have no quantitative features of an objective nature. They also have no causal role. The latter means that no conceivable scientific experiment could possibly prove or disprove the existence of the sensory qualities as physical entities. Therefore the conclusion that they exist in the physical realm cannot be subject to experimental verification but is one which can only be established, if at all, by reason. Also, because of their non-quantitative nature, whilst the addition of the sensory qualities to the physical world may make a substantial qualitative difference to our understanding of the universe it can make no *quantitative* difference. It follows that it can have no impact on the mathematical underpinnings or equations of the physical sciences. Therefore, while it is true to say that the thesis that relational sensory qualities exist in the universe represents an addition to physics, it is not one which has the slightest impact on either the mathematical theory or the experimental practice of that subject.

Their lack of objectively measurable features together with their nature as relationships also goes a long way to explaining the 'mysterious' nature of the sensory qualities. For in the current state of scientific knowledge colours, sounds, smells, etc. constitute the least understood entities in the universe. Not having any of the measurable characteristics of matter such as mass or electric charge they elude the grasp of contemporary science. Man has peered into the interior of the atom and contemplated events within 10^{-33} seconds of the beginning of the cosmos but still the very simplest features of the universe, its sensory qualities, do not yield to his understanding. We can see now that this is due to the fact that despite being physical they are not objects in the manner of atoms or rocks. Rather they are concrete relationships which span regions of space in the physical realm. I suggest that there are no other concrete relational entities in the universe, at least of a spatially extended nature, than the sensory qualities. If there were any then they would appear as if they were new types of sensory appearances.

DESCRIPTION AND RELATIONSHIP

Finally the concept of sensory qualities as relationships accounts also for, or at least is consistent with, one of their other more perplexing features. It is often remarked that man cannot describe the appearance of colours, sounds and smells or, at best, it is only possible for him to do so in terms of other colours, sounds and smells. But the same is true for many of the basic abstract relationships which occur in the physical arena. Thus length or weight can only be described by reference to arbitrary examples of those same relationships of length or weight, such as the metre sticks and kilogram masses maintained by various standards councils, which exemplify the very thing being described. The limitation of 'indescribability' therefore simply seems to be a characteristic of fundamental relationships in the physical realm.

10. A Relational Theory of Sensory Qualities

It is now apparent that colours, sounds, smells and so forth are not the simple entities of common thought. Rather they are relationships embedded in the arena of the physical world. The straightforward forms of direct realism that are based on the natural view must therefore be modified to take this into account. Here we will look at how this might be done.

Fortunately, as they are physical entities within the material world, relational sensory qualities have the objective status that is required to support the direct realist outlook. They exist alongside atoms and molecules without any dependence on being perceived. To that extent they have the correct set of characteristics for incorporation into this type of perceptual theory.

But building a 'relational' form of direct realism is not as simple as merely asserting that sensory qualities are relations within the physical realm which can provide an objective basis for perceptual experience. There are a number of further hurdles which have to be negotiated before the task may be declared complete.

OBJECTIVITY OF PRESENCE
The first hurdle concerns the status of presence across space: is presence across space a subjective phenomenon or an objective one? This question is particularly acute because it is noticeable that man universally experiences the characteristic of presence

across space as directed toward himself, the perceiver. For example, the blueness of the sky manifests downwards from the atmosphere, and it so happens that the human viewer is almost always located beneath the sky on the ground. Can this just be coincidence? Equally the yellowness of the sun is experienced as radiating towards the earth, which by chance is also the vantage point from where most human perceivers see it. So it is tempting to think that with sensory qualities the appearance of outwardness – or of manifestation 'towards' – is a result of the viewer having a perspective from which he or she becomes aware of those qualities. Presence across space, it might be argued, is not in fact inherent to sensory qualities themselves. The only thing that is inherent in our experience of them, it might be suggested, is the 'fromness' with which they are accessed by the perceiver. According to this way of thinking colour is exhaustively contained in the surfaces and volumes of objects and any sense that man has that it 'presents across space' is derived solely from the perspective which his visual processes impose on experience. On this view, 'presenting' does not persist in reality when a human act of perception ceases.

Such a view is opposed to the one offered here. I propose that the characteristic of presenting across space which sensory qualities display is a genuinely objective one. That is to say, it is a real feature of colour, sound, smell and so forth, which exists even in the absence of perceivers. On this 'objective view' the fact that sensory qualities are always experienced as targeted in the direction of the perceiver is merely a result of man's only being able to access such presenting qualities from a point of view. (It therefore has no greater significance than the similarly contingent fact that the eyes of a portrait painting may appear always to gaze at the viewer. They were not originally painted this way by the artist, and do not exist objectively on the canvas, as directed at each individual admirer of the piece.)

However, there is no means that I am aware of which can be used to prove that outwardness of presentation is an objective feature of sensory qualities. Although it might in theory be possible to build a proof by comparing the relative coherence of these two viewpoints regarding sensory qualities, I propose something more modest here. Namely, to judge simply whether the idea that presence across space is objective can produce a coherent picture at all.

So our next task will be to construct a form of direct realism that accepts presence across space as an objective feature of sensory qualities. Once this new theory of sensory qualities is built, at least in outline (towards the end of this chapter), we will take a moment to assess its degree of internal coherence.

PRESENTATION WITHOUT CAUSATION

In developing our account of the sensory qualities, the central concern will be that colours and other qualities present themselves across space. We want to merge this idea with direct realism, the concept that the world – and, in particular, its sensory qualities – are as they appear in perception. It will also be a working hypothesis, as supposed above, that presence across space is an objective feature of the sensory qualities. So whereas some existing forms of direct realism accept the reality of colour hues as residing on the surfaces of objects, and sounds and smells as being located in the environment, we will attempt to construct a version of the theory which accepts additionally the notion that qualities such as these present outwardly to points in the surrounding space – whether or not perceivers are present.

But what exactly might this 'presentation' or 'outward manifestation' involve? It cannot mean that little impressions of, for example, the colour red are made to exist at each point in neighbouring space. When the redness of a tomato presents to nearby locations it would be absurd to suggest that tiny images of its colour quality arise at each of them. The same holds true of the other sensory qualities. For presence of sound or smell across space it would be implausible to suggest that an aural or odorous quality occurs around each source event which is then replicated at every point in the surrounding space into which the quality is presented. Presentation of a quality to a point in space must instead refer to a more abstract type of event. Perhaps it refers to something more like the notion that access to, or availability of, the remote quality occurs at the point in space. In the case of the tomato this equates to each of its surrounding locations gaining – not a little image of redness – but rather accessibility to the redness quality located on the surface of the fruit.

What has to be borne in mind is that we are dealing here with *a relation*, albeit a concrete and physical one. As a connecting event, this is a type of entity that man is unused to contemplating within the context of the physical realm. It is quite unlike the rocks, planets, atoms and molecules which, together with their associated interactions, predominate there – at least numerically. It does not generate an event at its outer or 'destination' end in the way that causally linked series of events do. For relationships involving sensory qualities are of necessity entirely acausal.

We have already noted this in connection with the idea that the introduction of sensory qualities to the physical realm has no quantitative impact on the physical sciences. As appearances, sensory qualities such as the colour yellow and the smell of coffee cannot have any effects in the physical domain because they possess neither causal roles nor quantitative characteristics. A sensory quality 'presenting', then, is an event, or physical relationship, connecting two parts of the physical world but it is not a causal event. Thus the presentation at the 'destination' location is not caused by the source location. It occurs simply because the connecting event, which takes the form of a presenting across space, exists. The non-causal, purely existential nature of such connecting events or relationships may be thought of as rather similar in nature to the effects of space. Thus any two objects are not somehow *caused* to be kept apart by space but are so simply because the space between them exists.

It may seem anomalous to have such perceivable yet non-causal occurrences in the physical universe as those which sensory quality relationships represent. For, as we currently understand it, the matter of the universe is entirely causal in nature. Yet the occurrence of these relationships is an inevitable by-product of accepting that sensory appearances exist within the fabric of the physical world. Because such appearances are inherently non-causal, under direct realism we must get used to understanding events like the presentation of an appearance at a particular point as resulting only from the existence of a relationship to it – and not as the product of a sequence of causal events.

SENSE AND SIGNALABILITY

It is, however, known that particular wavelengths of light are associated with certain colours. Also, that there are similar correlations between frequency of sound waves and the pitch of the sounds that are heard, and between types of odour molecules and aromas. But these signals are causal in nature. So what connection could there be between them and the non-causal relationships of presentation across space which I have argued constitute all sensory qualities? It couldn't be the case that the signals simply cause the resulting sensory qualities, because that could only happen if the latter were the end products of conventional chains of causal events. Another possibility which suggests itself, however, is that the signals are involved in *determining* the qualities which are non-causally presented.

The best way to explore this possibility is to first examine some of the significant theoretical issues which it raises. We will then be in a position to apply the idea to concrete examples (including eventually a setting which incorporates the entire range of sensory qualities that are accessible by the human senses).

In the case of colour the associated signal is light, which is reflected (or emitted) from the surfaces of objects to each of the points in the surrounding space. In principle, therefore, it could be the case that light determines the hue of colour which is presented to each of those points. On this picture of the possible connection between physical signals and sensory quality relationships, other physical signals would carry out a similar role for their associated sensory qualities. So, instead of causing the quality of sound, sound waves would determine the pitch and volume of sound quality which is presented at each of the points that the waves pass through from a given source. Likewise the diffusion of molecules associated with olfactory and gustatory qualities would be held to be determining rather than causal factors in the presentation of those qualities.

LIGHT AND BLACK

One of the advantages of this way of looking at things is that it becomes possible, in the case of light and colour, to explain how the colour black is presented by entities when no light is emitted or reflected from them. Thus the depths of outer space and lumps of coal in a dimly lit coal-bunker both present blackness to their surroundings. This would be impossible to explain if the colour quality presented to a point of space was held to be the causal result of the light arriving at that point, for in these types of cases there is little or no such light. But instead we can now surmise that blackness is a default form of colour presentation. Because presence across space is a relationship rather than a causal event this does not pose any sort of a paradox. Blackness can be deemed simply the hue that is always non-causally presented by objects to points in surrounding space in the absence of light. Whereas when light is transmitted from the objects to those points it determines other hues to be presented (also in a non-causal fashion).

In Chapter 8 we envisaged that an object's colour can be presented as far into the cosmos as the light transmitted from it can travel without degradation of its information content through loss of energy. Again these extraordinary distances of presentation would be hard to accept if presence across space was a causal event, but they are rendered largely innocuous by the fact that it is acausal. But how far, then, may blackness be presented given that no light is involved here? The question is impossible to answer in any definitive sense, but it seems reasonable to suggest that as the underlying event of a non-causal presentation across space is identical to the case of other colours, then the range of presentation may also be similar.

EXPLANATIONS AND RELATIONS

So far we have surmised that causal events such as light and sound waves have a determining effect on their corresponding sensory qualities, and we have seen some of the benefits that this form of explanation can bring. But more precisely how *could* physical signals such as light and sound waves have an impact on non-causal connecting events of presence across space in such a way as to determine the quality which is presented?

The short answer again is that it is unlikely that such a question could ever be fully answered, or at least not at the level of microscopic mechanical detail that scientifically-aware modern man has come to expect in explanations of physical phenomena. The problem is that we come up here against the fact that the physical relationships which we call 'sensory qualities' are fundamental entities within the universe. Single hues of colour and individual sounds and scents have as qualities no internal elements or structure. This makes them arguably the simplest known entities in the universe. The result is that it is not possible to provide explanations in relation to these entities which are comparable in detail to those which modern science provides for more complex entities, such as those made of matter itself. However, it is possible to look for an account of the connection between causal signals and sensory qualities at a more abstract and generalised level. That is what I shall aim for here.

LIGHT AND INFORMATION

Having acknowledged that presentations of sensory qualities are relationships across spans of space, it seems probable that the only way in which they could be affected by causal signals such as light or sound waves is if, broadly, such signals modified the underlying relations that exist across those spans. This is indeed what such signals do.

For example, take light and colour. In the most general sense, two points of the universe are placed into a different relationship if they have light travelling between them than if they don't. If light travels to a location from a source object (residing, say, at point P) it places the former in a relationship with the latter of, roughly, 'being the recipient of a light signal transmitted from point P' (so these are abstract or conceptual relationships that we are talking about at this stage). Whereas if there is no light such a relationship between the two does not hold. Therefore, in this most general sense, light modifies the relations which obtain between the points in the universe from which it is transmitted and to which it travels.

But of course light has a number of characteristics and each of these has the potential to effect such relationships in more particular ways. For example, as a form of electromagnetic radiation, light may have high or low energy levels. So if the intensity of the light increases, the relation will take the form of 'being the recipient of a high-energy light signal transmitted from point P' while if it decreases the relation will take the form of 'being the recipient of a low-energy light signal transmitted from point P'. We know from human experience that brightness of apparent colour correlates with intensity of light. So I surmise that when the former type of relation obtains it determines colours to be presented at the destination location with high brightness. Whereas when the latter type of relation obtains it determines colours to be presented there with low brightness.

The How of Hue

There is another characteristic of light that may vary and that is its wavelength, a feature which is known to be especially relevant when it comes to explaining the hue of colour. When light is of a particular wavelength (call this W) then the relationship between destination location and source object becomes 'being the recipient of a light signal at wavelength W transmitted from point P'. (This encapsulates the fact that as the wavelength varies, the relation between the two points also changes.) As with intensity of light and the brightness of resulting colour, I propose that the nature of this relation at any given moment with regard to its wavelength determines the *hue* of the colour presented from a source object to a given destination location.

However, there is an important sense in which the wavelength of light is merely an embodiment of information which the light carries. In the case of sunlight reflected from the surface of an object, for example, its wavelength is an expression of information about the surface's physical micro-structure. Thus surfaces with a certain arrangement of crystals or molecules will cause light to reflect with a particular wavelength, while those with a different arrangement of particles will cause it to reflect primarily with another wavelength. (Likewise for light transmitted through translucent objects, wavelength is an expression of information concerning the objects' internal micro-physical properties of optical transmission.) The physical signals associated with each of the other sensory qualities can also be viewed as carriers of information about source objects or events (we will examine this in more detail shortly). So, this suggests that it makes sense to account in general for the link between signals and sensory qualities in terms of information. With this in mind I will emphasise in what follows that it is really the underlying *information* in light about the micro-structural features of objects such as their reflectance and transmission properties – given expression in the form of wavelength – which brings about the differences in relations that are of significance to hue of presented colour. As a consequence, the physical relationships of colour presentation between source objects and points of presentation are determined as regards to hue.

Having arrived at a relatively broad thesis about how the brightness and hue of the sensory quality of colour are determined by the causal factor of light, let us now apply it to some concrete examples and see how they work out in detail. Light reflected from green objects such as limes and blades of grass carries information about the source object's reflectance characteristics expressed in a predominance of wavelengths in a range centred on 550 nanometres. (This is often given the strictly inaccurate – because light is not coloured – but nevertheless convenient name of 'green light'.) On the other hand light reflected from red objects like tomatoes and strawberries carries such information in the form of a prevalence of wavelengths around 650 nanometres (so-called 'red light'). This means that a location (say, L1) in space receiving light from a red object, like a strawberry, is placed in a relation with it which might be described in simple (wavelength) terms as 'being the recipient of a light signal at wavelength 650 nanometres transmitted from S the strawberry'. Alternatively, this would be described in the more comprehensive, information-orientated terms, as 'being the recipient of light-borne information about S, expressed in a prevalence of wavelengths at *650* nanometres'.

A different point in space (call it L2) which receives light from a green object such as a blade of grass (G) has a relation with it – again expressed in terms of information – of 'being the recipient of light-borne information about G, expressed in a prevalence of wavelengths at *550* nanometres'. The difference in information content of the light arriving at L1 and L2 (or equivalently the difference in wavelengths, because the information content is expressed in the distinct wavelengths), is what gives rise to different relations across the spans of space between the strawberry and L1 as compared with that between the blade of grass and L2. My contention is that it is these different relations which account for the difference between redness being determined as presented from the strawberry to L1 and greenness from the blade of grass to L2.

Generalising from this it then becomes possible to understand more broadly how the wavelength of light transmitted from any object, which expresses information concerning its micro-physical attributes, can determine the colour presented to any point in the surrounding space through which it travels. As light, having been reflected from a surface with a certain wavelength, spreads across the environment, those points through which it passes have information-based relationships formed with the surface. As a result, the hue of the relational connecting event of colour presentation from the surface to those points is determined. In this way it becomes possible to envisage, at least in outline (which is all that can be hoped for given the ultimately simple nature of sensory qualities), how the capacity of light to bear information concerning the micro-properties of objects to their surroundings – no matter how distant[1] – determines the hues which are non-causally presented there.

INFORMATION AND SENSE

As we have remarked, the physical signals associated with the other sensory qualities also have information content. In a similar way, they too may have the capacity to modify underlying relations between points in space and 'source' objects or events. Thus sound waves carry information to the locations they travel through about impact events at their source location. These events may be anything from the collision of two solid objects (e.g. a hammer and the head of a nail) to the impact of a person's vibrating vocal chords on surrounding air molecules. In the case of sound waves, information about the energy and frequency of such impact events is expressed in the amplitude and frequency of the waves. The molecules associated with aromas and flavours equally carry information, in the form of their own molecular make-up, concerning the chemical composition of their source materials, to all locations through which they diffuse.

It therefore seems reasonable to suggest that for each of the types of sensory quality the associated physical signals determine the presented quality in essentially the same way as light does in the case of colour. That is, by affecting – through their information content – the abstract relations obtaining between source objects or events and 'destination' points in space.

SOUND WAVES AND SOUND

As an example let's take a detailed look at sound. A similar picture can be set out for sound as was given for the strawberry and the blade of grass in the case of colour. Take two tuning forks, one (Fork A) which gives a high-pitch noise and the other (Fork B) which produces a lower pitch sound. The sound waves that emanate from each transport information about the thousands of tiny impacts that the vibrating forks make on the air every second – their amplitude and frequency – to their surroundings. Fork A vibrates at a high frequency. That is, it makes a large number of impacts on the air every second and the sound waves rippling out from it transport this information, expressed in their matching high frequency, through the air to points in the surroundings. The sound waves pulsing out from Fork B do so with a lower frequency, expressing the information that it is making less frequent impacts on the air. Let us further imagine that fork B has been tapped more vigorously so that its impacts on the air have greater energy. This information is expressed in its sound waves through their greater amplitude.

Now consider a point in the surroundings (say P1) through which sound waves from Fork A pass. By analogy to the case of colour and light, its (abstract) relation to Fork A becomes that of 'being the recipient of sound wave borne information, expressed in frequency FA and amplitude MA' (where FA and MA are the frequency and amplitude respectively of the sound waves given off by fork A). Whereas another point (say P2) through which sound waves from Fork B pass gains a relation to it of 'being the recipient of sound wave borne information, expressed in frequency FB and amplitude MB' (where FB and MB are the frequency and amplitude of the sound waves emanating from Fork B).

This difference in abstract relations brought about by differences in the information content of the sound waves and expressed in their frequency and amplitude is what gives rise – if my reasoning in the case of colour and light is correct – to the difference in presented sound from Fork A to P1 and Fork B to P2. Thus the sound waves from Fork A transport the information that a high frequency of impacts per second are being made on the air (at a certain energy level) to surrounding points. As a result a high pitch sound (at a certain volume) is determined as presented from the vibrating fork to such points. The sound waves from Fork B transport the information that fewer impacts per second are being made on the air, although at a greater energy level, and so a low pitch sound at a higher volume is presented from it to the points its sound waves travel through.

MOLECULAR INFORMATION

Essentially the same idea works for smell and taste. As molecules from the surface of an aroma or flavour-giving object diffuse through a medium (air or liquid), they transport information (embodied in their own molecular constitution) regarding the chemical make up of the object from which they are derived. Although a cruder and more mechanical process is involved here than the others that have been considered there is no difference of principle. Thus the molecular-borne information modifies the relations between the points through which it passes and the location from which it originated. As a result it has the potential to impact on the aroma or flavour presented from the source item to those points.

INFORMATION AND QUALITY

There is a further question which needs to be asked at this stage about the precise nature of the possible connection between information and sensory qualities. Would *any* information-bearing signal or event that was conveyed between two locations result in the determination of some form of sensory quality presented between them?

The explanation of how causal signals determine presented qualities has relied heavily on the former's information-bearing capacity, and we know that there are many species of animal whose sense organs are sensitive to signals beyond our own. So this raises the possibility that perhaps the electric fields which catfish can sense, and also the ultrasound that bats use to detect their environment, are each linked with their own sensory quality which is determined by the associated physical signal. If that were the case then could the same could be true of a wide range of other physical phenomena (irrespective of whether or not organisms exist with sense organs capable of responding to them). This might include such a phenomena as magnetic fields and gravity which transport information about their source events.

Fortunately it is possible to say quite categorically that just because an event bears information regarding a source object to a location, it does not follow that this leads to a sensory quality being determined. This is shown by simple examples such as that of a conker (the nut of the horse-chestnut tree which is encased in a spiky outer skin) falling from a tree and landing on a leaf lying at the side of a river, where it leaves an impression of its needle-like spikes. The leaf, complete with a pattern of holes formed by the puncturing effect of the spikes, is then carried downstream by a surge of water and ends up caught further down the river by a boulder on the bank. As a result, information concerning the conker, in the form of the pattern of holes (not unlike the patterns of holes which convey information on a ticker-tape), has been transported from the source event of the fall all the way downstream to the destination. But it would be absurd to suggest that any sensory quality is determined as a consequence. So the transport of information concerning a source object or event may be a necessary but it certainly is not an adequate condition for determination of sensory qualities.

That being the case, what, it may be asked, is it that is special about light, which enables it specifically to have an effect on the sensory quality of colour? Equally, what is it about sound waves that allows them to impact on sound quality, and about certain types of molecules that enables them to determine the gustatory and olfactory qualities? After all, there are many other fields and forces that emanate from physical objects and which carry a certain amount of information about the micro-structure of those objects to the environment. Consider, for example, electromagnetic radiation in non-visible frequencies, such as radio-waves and micro-waves. Then there are also the magnetic and gravitational fields which penetrate the space around objects. Why don't either of these, by conveying information about an object to its surroundings, have an impact on the colour – or sound and smell – qualities presented from it? Under prevailing theories of perception it is easy, of course, to explain in a general sense why it is light that is associated with vision, sound waves with hearing and the diffusion of odour-carrying molecules in the air with smell. In each case these are the stimuli that correspond to the receptivity of the relevant human sense organ. As light is the form of electromagnetic radiation that stimulates the photoreceptor cells of the retina, inevitably it is this signal – rather than, say, gravity – which is associated with the sensory quality of colour that arises in vision. Sound waves trigger the auditory hair cells of the human ear so only this stimulus gives rise to the quality of sound, and so forth.

Matters are not so straightforward, however, when one is attempting to develop an account of the sensory qualities as independent of the percipient. Here one is presented with the challenge of explaining why it is that in the physical realm – in the absence of all perceivers and their sense organs – light rather than magnetism or gravity or whatever should be the single form of influence that gives rise to presented colour (and sound waves to sound, taste molecules to taste, and so forth). All one can do, ultimately, is point to the regularity of connection between such pairs in human experience and to the lack of evidence that other influences such as magnetic or gravitational fields have any direct impact on these sensory qualities. So any suggestion that colour depended on something other than light, or that the quality of sound was determined by a signal other than sound waves, would amount to nothing more than empty speculation.

THE HARD PROBLEM OF REALITY

At root both the direct realist and the more conventional quality-based representational view suffer the same ultimate inability to explain the link between sensory qualities and their associated causal triggers. While on the direct realist view we can surmise that light determines presented colour within the external physical realm, we cannot describe the detailed mechanics of how it does this. In a similar way, according to quality-based representational theories, it is impossible to say exactly how the specific activity of neurons within the brain gives rise to particular sensory qualities (redness, aroma of baking bread, sound of a guitar, and so forth) – hence, as we have seen, arises the so-called 'hard problem' of consciousness. In both cases these difficulties stem from the ultimately simple nature of the sensory qualities and the mismatch between this and the compound nature of the material realm as it is conceived within the physical sciences.

KITCHEN CONFIDENTIAL

Let us try and take an overview of the picture that has been developed so far in this new relational theory of sensory qualities.

To do so, imagine a kitchen with a vase of daffodils standing on a central wooden table. This is, let us say, a room of contemporary design with a large expanse of glass making up one wall and a low ceiling punctured with broad roof-lights that allow the sky to be readily visible above. As a result it is flooded with sunlight and the daffodils seem almost to glow bright yellow in the centre of the room. Nobody is present in the kitchen so this allows us to provide an account of the flowers' colour entirely in the absence of perceivers. We will eventually introduce onto the table a sequence of other objects in order to illustrate how each of the sensory qualities is to be accounted for under our new theory.

Firstly, then, the daffodils and the new theory's treatment of colour itself. Sunlight is reflected from the petals to almost every point in the space of the room. The only exceptions are beneath the table because its opaque surface blocks the passage of light, and a few other nooks and crannies such as behind cupboard doors. The reflected light carries with it a wealth of information about the microscopic cellular and atomic structure of the surfaces of the petals. As far as colour goes, however, the key feature of this information is that which is expressed in the light's mix of wavelengths. In this case that takes the form of a predominance of wavelengths at 580 nanometres ('yellow' light). The transport of this information places the points of space which the light reaches in a distinctive relationship with the petals' surfaces. As we noted in an earlier chapter it is not possible to describe relationships but, crudely, it might be possible to characterise such a relationship (from the point of view of the destination point) as something like 'being the recipient of light-borne information, expressed in a prevalence of wavelengths of 580 nanometres'.

The impact of this relationship is to determine a non-causal presenting across space of yellowness from each of the petals to all of those points in the space of the room through which the light passes and which thus have their relations with the petals' surfaces placed in this distinctive form. In fact, some reflected light also escapes through the windows and travels far through the external environment, so the yellowness is presented to points which are in certain cases considerably more distant than the confines of the room. There are also, we may imagine, daffodils growing in the garden outside, and light reflected from them has unimpeded access to travel up into the heights of the atmosphere as well as across the valley until blocked by a neighbouring hill. There must, of course, be a gradual depletion in the energy of light as distance from the surface of reflection increases and, correspondingly, also a loss of the information content it conveys. It follows that there is likely to be a similar degradation in the effectiveness of presence of colour across space.

TECHNICOLOUR

But now let us imagine that placed alongside the vase on the kitchen table there is also a blue tea-mug. The sunlight filling the room is reflected from its ceramic surface principally at 450 nanometres. The transport of information regarding its distinct surface characteristics – expressed in the 450 nanometre light – to the points of the room places their relations to its ceramic surface into one of 'being the recipient of light-borne information, expressed in a prevalence of wavelengths of 450 nanometres'. Hence blue is presented from its ceramic surface to them rather than yellow. Now consider a point of space close to the ceiling in what is approximately the centre of the room. This receives light from both the mug and the daffodils. As the recipient of light-borne information expressed at both 450 nanometres and 580 nanometres, it follows that it has both yellowness and blueness presented to it. There are of course many such points throughout the kitchen which receive light reflected from both the daffodils and the mug and so have blue presented to them from the item of crockery and yellow from the flowers. If presence of colour across space was a causal event this would represent a contradictory state of affairs, but as we have noted it is not an event of this type.

Instead, it is a *relation* and it is one of the characteristic features of relationships that entities can be involved in many of them simultaneously. This can be seen at the simplest level, from the fact that an object may, for example, at the same time be both 'bigger than' and 'to the left of' another. A town can be 'higher than', 'more populous than', 'wealthier than' and 'ecologically more sound than' another. This may be a feature unique to relationships and it is one which we will have cause to reflect on further in subsequent chapters. For it has considerable explanatory potential in the field of perception.

BLACKNESS AND INFINITY

Finally with regard to colour, let us consider what happens in the kitchen when there is no light. This could occur in one of two ways. Firstly, it could be daylight outside and all curtains and blinds could have been closed and the interior lights in the room kept off. In other words, darkness has descended but only inside the kitchen. Externally the sun still shines and objects continue to present their colours far into the environment. The daffodils in their vase and the blue mug remain on the table, so in such circumstances what does our theory say about their colour? According to our new theory, colour presentation is non-causal so it can continue from the flowers and the mug to all of the points in the room. However, because there is no light transmitted from them to such points the 'default' hue of black is presented under such circumstances. As we noted before no paradox ensues from the continuation of black colour presentation in the absence of light because the relationship of presence across space is not a of a causal nature.

We do not have a definitive concept of how far into the universe such blackness can be presented and this is evident when one considers the second way in which absence of light could arise in the kitchen – when night falls (without any effort being made to close curtains etc). In this case there would no longer be any sunlight to reflect off the surface of the daffodils or the blue mug so, on the current account they would again present the default hue of black to their surroundings. However, an observer looking in from outside (say from the garden), and conceivably at any distance away from the kitchen, would see the blackness presented from its interior. So we can conclude that the black colour presented by the flowers and mug in the absence of light may be presented over at least such a distance.

This raises the question of why in the first scenario the objects' blackness is not similarly presented to the realm beyond the confines of the kitchen. Of course, there are the brick walls that form the structure of the house between the interior of the kitchen and the garden. But if colour presentation is, as we have stated, non-causal it would be hard to explain how a wall of bricks, or any other physical impediment, could act as a barrier to it. One possible solution to this conundrum may lie in the idea that it is not so much that blackness is not presented to the external environment but more that this default presentation is *overridden* by all of the other colours which are presented there as a result of the light that floods the garden. Thus if we take a sample point of space in the garden, this might receive sunlight reflected (let us say) from the brick walls of the house and the curtains across its main windows. So any blackness that potentially could have been presented to this point from the objects in the unlit interior of the kitchen (and elsewhere in the house) would be overtaken by the colour presentation from the walls and the curtains brought about by their reflected light.

AROMATIC INFORMATIC

Once normal light is restored a thyme bush in a pot is added to the small collection of objects on the table. Within a short space of time the air in the room fills with molecules of volatile oils dispersing from tiny pores in the leaves of the aromatic herb from which they have evaporated. Each point at which they arrive has its relationship to the bush modified by the information concerning the chemistry of the bush's leaves embodied in the molecules. The fragrance which is presented from the bush to them as a result is determined as being that which we know of as the scent of 'thyme'.

A bowl of peaches arrives on the table next. Do these have a flavour? In fact, we have concluded that flavours are not inherent even in bodies such as these with a natural moistness. They only arise when the molecules of such an item disperse in a liquid medium – a process which normally requires a mildly enzymatic liquid like saliva in order to dislodge surface molecules (although it does not require a taster to be present). So the peaches as they stand in the fruit-bowl have no flavour. They also have furry exterior skins. Despite this they, and solid objects in general, do not have tactile textures in the sense of an external layer of 'feel' (although they may have the visual appearance of 'furriness'). For touch, we concluded, was associated with no sensory quality.

ENTER THE OBSERVER

The account of the sensory qualities which has been assembled here is internally consistent in the sense that all of the qualities are treated in the same way. They are each considered to be relationships of presentation across space which are determined by their associated causal signal. To that extent it is possible to conclude that the picture which has been developed is a coherent one. But what we have not yet addressed is how those qualities can become elements of experience to a perceiver. So what happens when a human being enters the kitchen which has so far acted as a conceptual laboratory for the development of our theory?

In the case of colour, the yellow of the daffodils continues to be presented (daylight having been restored) to every spot in the room which is reached by light reflected from their petals. So an observer moving into one of those positions will receive that same light as stimulation falling on his or her retinas, and one can presume that as a result the observer somehow becomes aware of the colour that is already presented to the location. (Thus as far as vision goes, the theory – being based on the transmission of light – predicts experience of the same colours at all of the same places as does conventional theory.) Much the same mechanism applies for the other sensory qualities. So, for example, the scent of the thyme bush is presented to every location in the room that molecules of its volatile oils reach. If a conscious perceiver were to walk around the room through such positions, so that his or her nasal passages are filled with air containing those molecules, then he or she would become aware of the presented scent. But can we say anything more about these types of processes? How exactly does a perceiver pick up or become aware of such pre-existing quality presentations?

The significant point here is that we have to think in rather different terms than we may have been used to under conventional representational theories of perception, in order to formulate such an explanation. For under our relational version of direct realism all of the phenomenology of the situation (its apparent colour, odour, sound and so forth) occurs in the external world. I suggest therefore that the only contribution which the observer's perceptual system has to make, in order to achieve what is called sensory 'experience' (or 'awareness'), is to supply an ability to achieve ongoing memory of the information that is embedded in the associated signal stream. This can be done through the conventionally understood human memory mechanisms that exist in the brain, and which are activated by nerve impulses flowing in from the sensory organs.

We will look at all of this in more detail in a later chapter. For the time being, I will talk in fairly simple terms of observers becoming aware of objectively presented sensory qualities, as a result of their sensory organs being stimulated by the associated signals and their neural memory being activated. But we can also note that as not only sensory organs but also neural memory structures deep in the brain are involved in processing the physical signals of light, sound waves and so forth transmitted from source objects, observers' entire perceptual systems participate in the information-based relationships with source objects. This will become significant later on when we examine some of the arguments most challenging to the direct realist position.

11. Problems

It has now proved possible to develop a direct realist theory of perception which embodies the relational nature of the sensory qualities. We must next test this new understanding of perception against the arguments conventionally held to disprove the direct realist position. This will require a thorough grasp of what those arguments involve, so that when we assess the ability of the new theory to solve them, we do so against a sound foundation. There are a wide range of such arguments (some of which we have encountered already) but they can be classified into a number of major categories.

EXTERNALLY CAUSED

First, let's take those arguments relating to perceptual phenomena which arise out of 'external' physical causes. In this category, for example, is the fact that when the path of light entering a person's eyes is bent by refraction, their vision of the world becomes distorted. Since Descartes pointed it out in the 17th century, philosophers' favoured example of this effect has been the 'bent-stick-in-water'. If you look from the side at a straight wooden stick that has been partially immersed at an angle in clear water, it appears broken or bent because the sunlight reflected from the stick is refracted at the boundary between the water and the air. As the stick itself is not physically bent, this is held to show that man's visual experience does not correspond to the actual arrangement of things in the physical world in the way that direct realism asserts. There are many similar examples of the distorting effect which refraction gives rise to, but a particularly obvious one occurs at swimming pools. Here straight lines of tiles at the bottom of the pool and the bodies of swimmers take on rippled appearances due to the passage of light through the waves on the surface. But neither the tiles nor the swimmers' bodies in fact physically fluctuate. So again what the viewer sees is notably different from the genuine composition of reality. This indicates that vision cannot be providing the kind of direct access to the physical realm which direct realism claims.

DIRTY WHITE VAN

The passage of light through a medium may not only have the effect of bending its trajectory, but can also filter out some wavelengths. This leads to discrepancies between an object's apparent colour and its 'true' colour (its colour when seen only through the medium of the air). For example, were you to view the daffodils of the previous chapter through the red glass lenses of rose-tinted spectacles they would look orange in colour because such glass only allows light to pass which is at the 'red' end of the yellow mix of wavelengths of light reflected from the flowers. This gives rise to an orange appearance. Yet daffodils are not orange, they are yellow. In the modern world coloured glass is widespread, and so, as a result, is this effect. Thus a worker in an office tower-block equipped with bronzed-glass windows, who gazes down at the cityscape below, will see the entire scene as imbued with an artificial brown-tint. For this reason when a gleaming white van stops at traffic lights below, it appears from such a perspective to be pale brown, as if coated in dust. The difference between the van's apparent and its real colour (visible when it is observed through an open window-vent by means of unfiltered light) again suggests that the direct realist's claim that vision provides immediate access to reality must be false.

NEWTON'S ROCK BAND

Differences between the colours which objects have, and those which they look as though they have, can also occur as a result of changing the wavelength of the light in which they are illuminated. As discussed earlier, Newton carried out a comprehensive study in which he demonstrated, as far as possible, that there were no types of objects whose apparent colour was not changed by such a procedure. But again in the modern world this is a phenomenon of everyday life. It is seen, for example, at rock concerts where the white T-shirts of band members change apparent colour from yellow to blue, then to orange and through the spectrum of colours, in co-ordination with the switching of stage lighting. (It is often only at the end of a concert when the house lights come up that the true colour of the musicians' clothing becomes visible.) The widely drawn conclusion from such effects – after Berkeley – is that because any object can be made to appear of virtually any colour 'there is no such thing as colour really inhering in external bodies.'[1] In consequence direct realism's claim that colour is an objective property of physical objects must be false.

INTERNALLY CAUSED

The next major category of arguments against direct realism concern those phenomena caused by internal factors involving the perceiver's central nervous system. The first and perhaps most obvious example of such arguments is known as the 'argument from illusion'. Philosophers can cite numerous forms of illusion in human perceptual experience which vividly reinforce the idea that in perception we cannot be directly perceiving reality. In the visual sense alone such illusory phenomena range from the full blown hallucinations caused by drugs like LSD, to what one might call 'micro-illusions' such as the after-images that are

143

experienced after staring for too long at powerful light-sources. It is also significant that illusions occur in all of our senses. Thus LSD, for example, causes gustatory and auditory hallucinations as well as purely visual ones. Prolonged use of cocaine can also lead to auditory illusions[2] and heavy users of the drug may even develop tactile hallucinations, such as the feeling that small insects have burrowed under their skin.[3] It is also possible to have the auditory equivalent of after-images, as occurs when you leave a loud rock concert with 'ringing in your ears'.

However, the philosophical point about illusions, in whatever sensory mode they occur, is that almost by definition what one perceives in them is not part of physical reality. Thus an acid-taker who sees green blood oozing from between the slabs of a pavement is experiencing something that is not actually there. Likewise a music fan who still hears the after-effects of a loud rock concert while making his way home is experiencing a sound that is not real. But if such appearances are not parts of physical reality then the only possible explanation for them would seem to be that they are anomalous elements which have arisen within the perceiver's mental representation of the external world. In other words, it appears to be an inevitable consequence of these illusions is that a quality-based representation is involved in perception and therefore that the process of perception is not direct. The same conclusion follows even from those micro-illusions of a visual nature which arise in the most mundane of circumstances. For nothing physical currently corresponds to the enduring yellow blob which continues as an after-image when you turn your sight away from a powerful lamp.

LOCKE'S BUCKET

The category of arguments based on the internal state of the sensory system extends beyond those based on the many illusions to be found in perception. It also includes, for example, a famous argument provided initially by John Locke. He pointed out that if you dip your hands into a bucket of tepid water but one of them is cold while the other is warm, the water will feel hot to the cold hand and cool to the warm hand. But 'it is impossible,' as he said, 'that the same water should at the same time be both hot and cold.'[4] This strange phenomenon seems to indicate that the sensory quality of heat – and perhaps, by extension, all sensory qualities – is not intrinsically possessed by matter in the way that direct realism requires. Rather it may have some dependence on the state of the perceiving system (in this case the skin and its temperature sensing organs) of the person detecting it.

ANIMALS AND APPEARANCE

Many animals are thought to experience the world differently to us. The resulting perceptual conflicts have always been presented as a considerable challenge to direct realism. Because they arise from the fact that animals' central nervous systems – including sense organs – are different to humans, this group of arguments also fall into the category of arguments against direct realism resulting from 'internal' factors. The existence of distinct forms of animal vision firstly undermines the idea (which is implicit to direct realism) that any object under given lighting conditions has a single definite colour. For if animals see the object (or would see it if they were present) as being a different colour than man then what basis can there be for any claim that the colour apparent to humans is the 'correct' one? After all we are only another kind of primate and even amongst humans there may be a certain amount of variation in colour experience due to degrees of colour blindness within the population.

The majority of arguments against direct realism take as their starting point situations (which are surprisingly common) in which there is a discrepancy between how objects or fields of view appear in perception and how it is considered that they in fact are 'in reality' at that moment. From the existence of this difference, the conclusion is drawn that perception does not provide man with an opening directly onto reality. But the arguments associated with animal perception do not follow this pattern. Instead they point to alleged inconsistencies or impossibilities within the direct realist position, as demonstrated by the existence of animal perception, and draw the conclusion that direct realism must be false. This can be seen clearly in a second argument associated with animal perception.

There are present in nature large numbers of animals with visual systems which are sensitive to different ranges of light than the human. All such perceivers may potentially see identical objects as possessing distinct colours. For example, the bee's eye responds to ultraviolet, blue and green but not to red wavelengths of light, giving it in all probability a quite different visual experience of the world to the one man is familiar with. Animal consciousness is a contentious subject and not one we can get sidetracked into examining in depth here, but for the sake of the current argument let us make the plausible assumption that bees have at least a form of visual colour consciousness. In that case, it has been proposed[5] that a bee's unusual range of visual responsiveness would cause it to see many aspects of nature in dramatically different colours from those apparent to humans. For example, the bee's ultraviolet sensitivity would reveal to it certain patterns that are invisible to humans; such as a dark centre on the lesser celandine flower whose petals appear uniformly yellow to man's eyes. There are also numerous instances in nature of this same type of conflict between the animal and human view of the world. These include sheepdogs with their famed ability to hear sounds at pitches higher than those that we can detect, and, extraordinarily, the goldfish which has eyes so sensitive that their responsiveness extends all the way through the spectrum of light from ultraviolet to infra-red. Such a feat makes the visual receptivity of this innocuous-looking fish wider than almost any other animal on the planet.[6] Given the opportunity, a goldfish would be able to see not only the same pattern on the lesser celandine as bees but also the beams of infra-red radiation used by television remote controllers in many modern households.[7] So if there were both a bee and a goldfish sharing a house with its human occupant, it is clear that these animals might perceive many parts of the interior as possessing colours which conflict with those experienced by the homeowner. Because of the existence of colour-blindness the same problem, albeit in a less dramatic form, can occur solely between humans. Thus if two people are viewing a tomato, and one of them has red-green colour-blindness, then the colour quality which he or she experiences as on the fruit may differ to that which is apparent to the viewer with 'standard' colour vision.

But if, as direct realism claims, colours are inherent properties of objects then it should not be possible for multiple colour qualities to occur simultaneously on individual objects. So, it is argued, the existence of animal perception, and to a lesser extent colour blindness amongst humans, shows that the colour qualities we see cannot be inherent properties of the objects that they appear to be located on. The only way of accounting for the multiple colour appearances of different classes of viewers would be if those appearances were products of mental representations within the perceivers' brains – i.e. by the falsity of direct realism.

GALAXY IN MOTION

A final category of arguments against direct realism involves those derived from contemporary scientific knowledge. The first argument of this kind that we will examine is as follows.

Our Milky Way galaxy rotates, like other spiral galaxies. This means that all of the stars in its spiral arms are in constant motion as they follow their broadly circular orbits around its centre. The sun, together with its solar system of planets, is located towards the periphery of the galaxy and is known to orbit the galactic centre at a speed of 250 kilometres per second[8] (this means that it takes 200 million years to make every complete circuit). Other stars have somewhat different orbital speeds – their speed is slower, closer to the centre – but they are all of a similar order of magnitude and stars at a comparable distance from the centre as the sun have a correspondingly similar speed.

So what is it that you are actually seeing when you look upwards into the night-sky and observe one of the multitude of stars in the star-fields of the Milky Way? Let us say that the star under observation is also located close to a point near the perimeter of the galaxy and so has a rotational velocity around the galactic centre of approximately 100 kilometres per second (a figure which is of the same order of magnitude as that of our own sun). The distance from our sun to the centre of the galaxy is 26,000 light-years and the thickness of the galaxy at its position is 2,000 light-years, so we can estimate that the star being viewed must be within those distances. A reasonable figure for the sake of the present argument might be that it is 1,000 light-years from us here on earth.

Light, travelling at its finite speed of 2.998×10^8 m/s takes one year to travel every light-year of distance. So the light which is presently arriving at the earth from the star, and which enables you the observer to see the little white point in the night-sky which tells of the star's existence, must have been emitted 1,000 years ago. But if the physical star itself is travelling 100 kilometres every second through galactic space, how far has it moved from the position at which the light was emitted during the intervening 1,000 years? Clearly, many millions of kilometres (in fact, approximately 2×10^{10} kilometres). So the position at which an observer sees the white point now – the position from where its white colour is presented – contains only empty space. But this tells us that the visible colour of the star cannot truly be located out in the physical universe where it appears to be. For it would mean that the apparent whiteness of such a star – and by similar arguments, the whiteness of many other stars in the Milky Way – was 'floating' in empty space.

This is a 'time lag' argument, similar to another based on the fact that a number of the 'stars' visible in the night sky may physically no longer exist, having exploded as supernovae thousands or even millions of years ago. But due to its finite speed of travel, light currently arriving at the earth continues to be derived from the pre-explosion stars and so gives rise to the ongoing appearance of luminous white bodies. In their case also it would seem that if the visible white colour we experience were genuinely presented from inter-stellar space, it would have to exist in space that was empty of all matter. In general 'time lag' arguments make a strong case that much that one thinks is seen directly in the physical realm in fact cannot be.

THE SPACE OF MATTER

Modern physics has also revealed that even matter itself consists primarily of space. This startling fact arises from scientists' increasingly accurate appreciation of the size of the constituent particles of atoms, of which matter is ultimately made. The dimensions of such particles as neutrons, protons and electrons have turned out to be minute relative to the scale of the atoms within which they exist. As a result, it has become apparent that most of the volume of each atom consists of the space between its component particles. To take the gold atom as a not-untypical example, its central nucleus of neutrons and protons has a radius that is well under 10^{-4} that of the atom as a whole.[9] The rest of the atom is made up of space and a number of electrons orbiting around this minuscule central point. The electrons are so tiny that 'they are as close to being point particles as present measurements can establish.'[10] In other words far and away the bulk of the gold atom consists of empty space. Similar proportions apply to other atoms. As all matter is atomic this indicates that matter in general consists of a high percentage of space and very little of what has traditionally been thought of as 'solid stuff'.

But if material objects consist largely of space then it follows that there simply is no 'solid stuff' within the micro-structure of matter onto which colours as sensory qualities could be anchored. It is hard, therefore, to see in principle how colours – or sensory qualities of any types – could be objectively owned properties of matter.

Taken individually, each of these philosophical arguments provides a powerful case against direct realism. In so doing, they also provide impressive backing for the position at the heart of the conventional, Newton-inspired picture of perception, that there is a quality-laden representation which mediates between the observer and reality during acts of perception. But the fact that there is such a broad array of arguments all pointing to the same conclusion makes it seem almost unthinkable that the representational position could be incorrect.

12. Solutions

Next we will examine whether our new relational theory of sensory qualities can resolve the powerful arguments against the direct realist outlook. Over the next few chapters, I will consider each major category of argument in turn. Let us first consider the arguments involving *external* physical conditions such as the effects of coloured filters.

KITCHEN CONFIDENTIAL II (THE SEQUEL)

Let us return to the uninhabited kitchen that acted as a conceptual 'laboratory' for the initial development of the relational form of direct realism, where a vase of daffodils sits on a long central wooden table. But this time we shall take it that an additional item has been set up next to the vase. It is a tall panel of glass mounted in a slotted wooden stand. The sheet of glass reaches most of the way up to the fairly low ceiling of the room and because it is only loosely mounted in the stand can be exchanged for glass panels of other types, of which there is a selection leaning against a wall.

With a transparent glass sheet in place in the stand to begin with, sunlight reflected from the daffodils is unmodified as it passes through the panel. Therefore the way that the flowers' yellow colour is presented to the room remains as it was described in Chapter 10. However, now let us imagine that a panel of pale blue glass is put in its place.

Glass of such colour has the effect of filtering out all but shorter wavelengths (that is, blues and greens) from any light that is transmitted through it. Of the 'yellow' mix of sunlight reflected from the daffodils, with a peak at around 580 nanometres, now only shorter wavelengths can pass through to reach points in the room beyond. A yellow object seen through such a blue filter looks green, so the wavelengths which pass through the blue glass from the daffodils will be those at around 540 nanometres (corresponding to 'green' light). Now only light having a wavelength at around 540 nanometres will arrive at locations in the kitchen on the far side of the pane from the flowers. Relational colour, we have argued, is determined by the information content of light about a source object, as expressed in its wavelength, passing to the point of presentation. But as only wavelengths of 540 nanometres – those correlated with the colour green – and their associated information, are able to pass through the blue glass filter, this means that green will now be presented from the daffodils' petals to the affected parts of the room. Yet at the same time yellow continues to be presented to all other parts of the kitchen, because they still receive unmodified yellow light at around 580 nanometres from the flowers which has not passed through the pane. The tall blue panel of glass reaches close to the ceiling, so positions in half of the kitchen are presented with green from the daffodils while those in the other half (receiving unfiltered light reflected from the flowers) are presented with yellow.

Note that no observers have been introduced into the room as yet. According to our new theory all of this occurs objectively and quite independently of the presence or absence of any persons.

THE OBJECTIVITY OF THE OBSERVER

But let us now supply the scenario with the one thing that it currently lacks – an observer. Imagine that you decide to refresh the water in the daffodils' vase. So, you enter the kitchen, moving into the half of the room which is receiving green filtered light from the flowers. Looking towards the daffodils through the blue panel of glass, what would you perceive? Because of the green-wavelength light stimulating the retinas of your eyes you would see what are 'really' yellow flowers looking green. The conventional argument uses the discrepancy between this apparent and the 'real' (yellow) colour of the flowers to argue that the appearance can only be accounted for by the existence of a mediating representation underlying your perception. This viewpoint concludes that vision does not provide man with direct access to the physical world. But the relational conception offers us here an entirely different account. What it says is that the so-called 'apparent' green colour is actually objective and real. For that is the colour which the daffodil presented to the half of the room that you now occupy before you arrived there, and will continue to present after your departure. It is neither your mind nor a representation within it which has produced the green colour. That colour is presented to your location whether or not you are there.

Remarkably, the relational picture of colours succeeds in accounting for the coloured appearance that objects have when seen through coloured media without any necessity to invoke the belief that it has arisen in the mind of the perceiver. The apparent colour in such cases is no mere subjective appearance. Rather it is objectively real and is simply experienced directly by the observer.

ROSE-TINTED WORLD

The principles of this explanation extend to a wide range of real-world conditions of perception – in fact any situation where objects are viewed through coloured media such as glass or Perspex and appear to take on a different colour as a result. The example cited in the previous chapter – of a white van looking dirty brown in colour when viewed through a certain type of window – is a case in point. Here the broad mix of sunlight reflected from the side of the van is filtered down to a range equivalent to murky brown by bronzed office windows. It is for this reason that such a van presents not whiteness but pale brownness into the interior of offices equipped with this type of window. But it does so even when they are unoccupied. The brown colour is not in the minds of those who work inside the offices.

The same principles of explanation also provide a ready account for the effects of rose-tinted glasses (or spectacles of any other hue). Consider a white marble sculpture when seen through spectacles of this kind. Is its pink appearance an artefact in the viewer's mind? An object such as a white marble sculpture reflects a broad mix of sunlight ('white' light). But only longer wavelengths of light (reds and oranges) are able to pass through the glass of red-lensed spectacles. So the mixed light reflected from the sculpture is filtered down to light containing a band of wavelengths which predominate in the 'red' region. This is equivalent to the mix of light which gives rise to an appearance of pale red or pink colouring.

As a result, our new relational theory of sensory qualities explains that an object like the white marble sculpture presents the colour pink through spectacles such as these – whether anyone is wearing them or not. If they happen to be donned by someone who directs his or her view toward the sculpture then he or she becomes aware of the pinkness which the marble objectively presents through the lenses. (Meanwhile the marble presents white to the rest of the world, where its reflected light remains unfiltered and of broadly mixed wavelength.) Once again the person's colour experience is the 'correct' one in the sense that it is the one which the marble presents through the lenses even if the spectacles are not being worn. While at the same time the object continues to present its 'normal' colour (in this case white) to other parts of the world – and to any observers located there, who receive light directly without it passing through the coloured filter. The colour observed through the spectacles is an objectively presented one and there is therefore no need to invoke a quality-based perceptual representation to account for it.

These examples demonstrate that the relational nature of colour, as embodied in our new theory, undermines the force of the argument derived from the effects of coloured filters. It is no longer necessary to conclude from experiences gained by looking through coloured filters that a mediating quality-based representation must exist within perception, nor that direct realism is untenable. One begins to sense that the blending of the relational concept with direct realism may have the potential to provide dramatically new ways of responding to long-standing arguments against the idea that sensory qualities are objective. The question is whether it provides the resources to address all such arguments with the same accomplishment. The answer to that will be revealed over the next few chapters.

13. Colour and Newton's Experiment

How does our new theory cope with the argument based on the fact that the colour appearance of objects changes as the wavelength of the ambient light is adjusted? This occurs, for example, when white paper is moved from daylight to the illumination cast by a coloured lamp. The paper appears to take on the hue of light that shines upon it. The phenomenon was first documented by Isaac Newton in an experiment in which he showed that every type of object which he could obtain – from paper and ashes to gold and peacock's feathers – changed apparent colour when illuminated under coloured light. He subsequently commented in a famous 'definition' regarding the nature of colour that 'Colours in the Object are nothing but a Disposition to reflect this or that sort of Rays more copiously than the rest'.[1] However, as noted in an earlier chapter, it was Bishop George Berkeley who brought to light the philosophical core of the observation by arguing that it showed that 'there is no such thing as colour really inhering in external bodies'.[2] But if colour qualities are not inherent 'in' objects then it seems to follow that, contrary to the dictates of direct realism, they must be in the mind of the viewer.

To assess whether the relational theory can deal with this phenomenon, let us recall first how the normal white colour of a sheet of paper in the example above is explained. In daylight the paper reflects to its surroundings a broad mixture of wavelengths from the ambient sunlight. In these circumstances, mediated by such mixed wavelength light arriving at each point in the volume of space around it, the paper presents white to these points. Next consider what happens when the sheet of paper is moved into light cast from, say, a blue lamp (or in Newton's case blue light split out from the spectrum of sunlight by a prism). The ambient light flooding the paper now contains exclusively light of 450 nanometre wavelengths ('blue' light) so the paper can reflect only these to its surroundings. Mediated solely by blue wavelength light the paper as a result presents blue colour to locations around it rather than white. The key point here, however, is that, as much as with its previous whiteness, such presentation of blue colour occurs objectively (i.e. even in the absence of observers). When a person sees the sheet as blue therefore this is not an appearance which in any way differs from reality. On the contrary, it more correctly represents the paper's appearance under the conditions of its being illuminated by blue light than would seeing it as white. Although the sheet of paper is nominally described as 'white', according to our new theory its colour under a blue light *actually* and *objectively* changes to blue. If that is the colour that a person sees it as possessing in these circumstances, then once again he or she is seeing it directly and unaided by a mental representation.

The principles of this explanation apply broadly. So when a rock band's 'yellow' T-shirts appear orange under coloured stage lighting the same thing is happening. As a result of being illuminated solely by red overhead lights they can reflect only an orange mix of wavelengths back into the environment. This determines an orange colour to be presented at each point in space through which the reflected light passes, rather than the normal yellow. But the orange colour has not been created in the mind of each audience member. It would continue to be presented to the space of the concert hall even if the band played to an empty auditorium.

In, Out or In-between?

So what, then, did Newton's experiment truly show concerning the objectivity of colours? Ironically, both Newton and Berkeley may have been correct to conclude from it that colour is not 'in' objects. But where the line has been overstepped is in the subsequent deduction by some thinkers (motivated by Newton's depiction of vision in *The Opticks)* that colours *must* be 'sensations' in the perceiver's mind. On the contrary, they could be not 'in' objects by being relations which span the gap from those objects to the surrounding space. The evidence which has been put forward during this and previous chapters makes the case that they are indeed exactly such relations.

We have examined so far two of the principal arguments against direct realism within the category of those arising from 'external' physical causes. In both cases, the new idea that sensory qualities are relations has delivered everything that might have been hoped for. It has shown itself able to explain the phenomena in question without any need to resort to the inner perceptual representation on which the rival representational model of perception depends. The door to a direct realist world view has thus been inched open a fraction. There remain, however, many profoundly challenging arguments to be faced by this new relational theory of sensory qualities before that door can be swung open further.

14. The Bending of Light

The next group of arguments against the direct realist position in the 'external' category which we need to consider are those based on the refraction of light. The distorting effects of refraction arise frequently in daily life. Two examples that we have so far considered involve the 'bent' appearance of a stick in water, along with the optical illusions caused by water in a swimming pool (where straight lines of the tiles appear 'wavy'). The question is: can the new theory account for these discrepancies between reality and appearance?

REFRACTION AND REALITY

First let us concentrate our inquiry around the case of the apparently bent stick-in-water. Assuming that the stick is made of wood, it reflects a complex mixture of sunlight which corresponds to the colour that we know of as brown. Prior to its being dipped in the water this light is transmitted straightforwardly to all points in the surrounding space. Hence, according to the new relational way of thinking, its brown colour is presented to all of those points. But when partially immersed in a pond or river, the refraction of light at the water-air boundary introduces a complicating factor to the way that the stick's colour is presented to its surroundings. The wavelength mix of the reflected light is not affected, so the apparent colour of the stick does not change, but the trajectory of the light is altered and therefore the colour of the immersed parts of the stick is presented to different locations than before. Once the stick is placed in the water, light reflected from the lower, underwater portion of the shaft travels along paths that are bent by refraction when they pass through the water's surface. Because of the resulting change in trajectory, there are points in the surrounding space which receive light arriving from the immersed, lower part of the shaft which would not have done so had the stick been resting at the same angle in air (and thus all light rays from it had followed straight paths). The refracted light from the immersed parts of the stick arriving at those points causes brown to be presented to them. But at the same time brown is presented from the upper half of the stick to points in the surrounding volume of space, mediated by rays of light which have taken straight paths because they have not passed through water and been refracted at its surface. Thus the colour brown is presented from the lower half of the stick to points which receive rays which, due to refraction, have taken bent paths. Yet it is presented to points from the upper half which receive rays that have taken straight paths. The points to which the brown from the lower half is presented therefore do not line up with those to which the brown from the upper half is presented. Thus arises the appearance of a bent 'stick-in-water'.

Once again, the significance of this explanation is that the appearance occurs *whether or not there are observers perceiving it*. The stick itself, of course, remains physically straight throughout the event, and once it is partially immersed in water its colour appearance presented to all points in the surrounding space, becomes *objectively* that of a bent brown stick. That is to say, it is objectively the case that the points in surrounding space to which brown from the upper portion of the stick is presented do not line up with those to which brown from the lower portion is presented. In seeing the stick as 'bent' therefore a person perceives it accurately and quite possibly directly. There is no necessity to invoke the idea of a mental representation to account for a discrepancy with the stick's objective appearance, because there is no such discrepancy.

THE PHILOSOPHER AT THE SWIMMING-POOL

These principles of explanation extend to all examples of refraction in vision – whether they occur as a result of looking out of bathroom windows made of rippled glass or through the disturbed surface of water at a swimming-pool. To illustrate, let us consider the case of the underwater lane markings at swimming-pools which appear 'wavy' due to the action of water on the pool's surface. First, take the situation when the water is flat calm. (We will also assume that there is no-one in the pool hall, so that our account is given – initially at least – entirely without the involvement of observers.) In this instance, each line of dark tiles presents a perfectly straight appearance to the surroundings because light rays reflected from them are undisturbed as they travel through the water and into the air. (Although refraction does occur at the water-air interface, it is to a constant degree and so causes a uniform displacement of apparent position rather than any distortion of shape.)

Now imagine that a single wave travels along the length of the pool over its otherwise perfect surface, perhaps due to a brief surge in the water supply. Let us consider the effects of this on how the dark (blue, say) colour of the tiles is presented to the empty space of the pool hall above the water. In particular, how it is presented to a single randomly-selected point in the hall, say two metres above the walkway on the right-hand side of the pool.

When the surface of the water was flat and calm, all of the light rays arriving at this point from the row of dark-blue tiles followed straight paths[1] and so the colour dark-blue was presented to it uniformly from the entire length of the line of tiles. But now the wave acts as a 'hump' of water travelling down the length of the pool. This means that at any given moment, the light rays arriving at the selected point from a short section of the line of tiles will be undergoing a complicated series of twists as they pass through water-air boundaries on both the rear and front faces of the hump. As a result, the dark blue colour will be presented to our selected point in the pool hall from places that are out of line with those from which dark blue is being presented by the rest of the tiles (whose light rays travel to the point through calm water). The consequence is that the overall appearance which is presented to our selected point above the right-hand side of the pool is that of a largely straight line of dark blue tiles but with a kink in the middle. The location of that kink moves along the apparent line of tiles as the wave progresses down the pool. So it is that the appearance of a kink in a line of tiles arises as a wave travels over the water.

The colour appearance of a straight row of tiles with a small section or 'kink' off-set in the middle of it is not, of course, just presented to our randomly selected position on the right-hand-side of the pool. It is also presented to an equivalent location on the left-hand side and many other points within the space of the pool hall – indeed every point that light rays reflected from the line of tiles arrive at which have had their trajectories twisted by passing through the wave. Furthermore, in a busy, real-life swimming-pool this effect is multiplied and complicated many times by the hundreds of waves travelling across the water's surface at conflicting angles, so giving rise to the shifting and 'rippled' appearance of underwater features such as swimmers bodies and tiled lane markings. But, as waves pass across the surface of the water and twist the path of light rays, this presentation of colour to points in the pool hall – from positions that are off-set with respect to the location of physical features underwater – is objective and occurs whether or not observers are present. When an observer moves to occupy a point, and looks into the pool, they merely become aware of the colour qualities that are objectively presented there as a consequence of their eyes being stimulated by the pattern of light signals arriving at the point.

15. A Bee's Eye View

The question of whether animals are conscious of sensory qualities in anything like the way that humans are is a matter of considerable debate in contemporary philosophy. If animals are, then significant problems are raised for the direct realist position. In this chapter we will examine those problems and ask whether the relational concept can offer a means of resolving them. For the purpose of this work, I will not engage in the debate itself, but will simply assume that animals do enjoy a straightforward form of sensory awareness. There are two reasons for adopting this approach. Firstly, it would be a diversion from the main thrust of the work to digress into the field of animal consciousness. Secondly, by assuming the worst case for the theory advocated here, we force ourselves to respond to the arguments which follow from it.

In order to fully draw out the case against direct realism, I will take it that animals possess a form of sensory consciousness which in its broadest characteristics is not unlike man's. In other words, that in acts of perception they become aware of sensory qualities in a largely similar way to that in which humans do. Of course, the sensory qualities themselves which animals detect may be different from those sensed by humans but the form of awareness they enjoy I will take to be essentially the same. In the future, when researchers have built up a full picture of animal sensory mechanisms (and perhaps even derived a coherent account of the evolution of animal sensing for the entire animal kingdom) it may turn out that the assumptions I make in the present chapter do not hold strictly true. But that should not significantly affect any of my theses here, because their purpose is to address animal perception *in principle* and to explore in general terms how the relational concept might overcome the difficulties which it presents.

Given the assumption that animals have sensory consciousness, the main problem[1] that is raised for direct realism derives from the fact that animals may experience conflicting sensory qualities to humans, as a consequence of their sensitivity to distinct ranges of sensory stimulation. For how can colours be inherent properties of objects that are perceived directly by viewers, if humans experience objects as having one colour and animals another? If, for example, a flower is seen as being red by man and purple by a species of insect?

For the red colour quality of, say, a rose to be intrinsic to its petals, then under similar conditions all viewers who see the flower *directly* should experience it as having that unique colour. It seems we can only account for the fact that some animals see the rose's colour differently to us by saying that their experience is the product of a mental representation, and that they therefore do *not* see the flower directly. But then, given the evolutionary continuum between animals and man, it would be hard to sustain the concept that human vision is direct. For, at best, it would be implausible to suggest that man alone in the animal kingdom enjoyed direct visual access to the colour qualities of the physical realm. In broad physiological terms, the underlying mechanisms of all animal and human visual systems are similar, so there would be little biological basis on which to make such a radical distinction.

EYE OF A BEE

In Chapter 11 we noted the bee and the goldfish as two examples of animal species with extreme non-human visual sensing capabilities. Here we will look in some detail again at the bee. Because it possesses compound eyes, the entire visual world of the bee must be wholly different from that of man's.[2] In terms of colour awareness the major distinction, however, between the vision of humans and that of bees, derives from the bee's sensitivity to light of different wavelengths. The retina of the human eye has three photo-pigments which respond to blue, green and red wavelengths of light. In practice, most instances of colour experience involve stimulation by a mix of these wavelengths and thus the activation of all three of the pigments. The eye of the bee, on the other hand, is sensitive to ultraviolet light. This comes about because although like man it has three pigments, these respond to blue, green and *ultraviolet* wavelengths of light.[3]

Because as yet there exists no comprehensive, evolutionarily-based account of animal sensory consciousness, we cannot be certain what the visual experience of the bee is like. However, we can at least imagine that the sensory qualities which animals experience are linked to the sensitivity of their sense organs, in the same way that the sensory qualities which man experiences are linked to the sensitivity of his sense organs. In other words, taking the case of vision, it is a plausible assumption that whenever an animal's eyes are stimulated by light exclusively composed of a single wavelength that corresponds to the sensitivity of one particular eye pigment, the animal always sees the same hue of colour quality (just as is the case for man). Such an assumption allows us at least to deduce what the degree of variation is in an animal's experience of sensory qualities, from an understanding of the sensitivity of its sensory organs, and to compare this with man's. Thus the bee has three photo-sensitive pigments in its eye just as man does. On the basis of this assumption we can conclude that it enjoys the same range of variability in primary hues as man. However, as one of those pigments provides sensitivity to ultra-violet wavelengths of light – a feature not present in the human retina – the nature of the colour qualities may in some and perhaps many cases (given that most real-life objects reflect at least a proportion of ultraviolet light) be different. In the simplest case, when pure ultraviolet light is given off by an object then the bee may see it as possessing a colour quality totally unknown to man.

To Be a Bee

It is a big leap to place ourselves, as it were, inside the head of a bee, and to imagine what its colour experience of the world is like. For example, the bee has one pigment which like ours is sensitive to blue wavelengths of light. So can we say that where there is a pure source of 450 nanometres ('blue') light – such as the sky – a bee will see it as possessing the same hue as a human does? One person who implicitly suggests that we are justified in making this assertion is zoologist and wildlife photographer (and producer of the television series *Supersense: Perception in the Animal World*) John Downer. He proposes that if a human could see the world with eyes adjusted to the sensitivity of a bee's then he or she would still see the sky as 'reassuringly blue with fluffy white clouds'.[4] But virtually everything else would look completely different. Thus 'previously green plants would be transformed into a bewildering array of colours. The grass might appear red, some trees would become magenta, and against a light background red flowers, such as poppies, would stand out as black blooms.'[5] I do not suggest that this gives a fully accurate picture of the true nature of the chromatic experience enjoyed by bees. Nor, I imagine, did Downer intend it to be taken as such. But it at least gives us an idea of the extent to which – given the various assumptions just made – the bee's colour experience of common objects such as flowers and trees must contrast with our own. It is this difference in apparent colour for identical objects that gives rise to so much difficulty for direct realism.

Although we do not currently have the means to verify the specifics of the colour qualities that bees experience, we do know of one visual quality which they are *likely* to experience in a different way to man: any colour quality that may be associated with ultraviolet light. This is because of the ability of the bee's eye, unlike that of humans, to respond to such light. It is conceivable that flowers and other objects which reflect ultraviolet light may present an associated 'ultraviolet' colour quality to the surroundings. If so, with their ability to respond to this wavelength of light, bees may be able to access the quality – but humans certainly cannot. One such flower is the lesser celandine which appears uniformly yellow to man but which, because it reflects ultraviolet light from a central region of the flower head, may appear to bees as having a distinct visual quality in this region.

Whether or not it is true, as Downer suggests, that bees see grass as red and poppies as black, there is a significant likelihood that they see these common plants as something other than the green and red which man sees them as respectively. But there is also a strong case that regions of flowers like lesser celandines, which reflect primarily ultraviolet light, appear to bees with a visual quality that is distinct to that which is experienced by man. It may even be a colour quality that can never be accessed by humans. Whatever the details of the final analysis, it seems highly likely that there are many objects in a typical garden or area of countryside that, if simultaneously viewed by a bee and a human, would have different colour appearances. As we have seen, this gives rise to a major problem for any theory based on direct realism. For if the uniform yellow seen by humans on, for example, lesser celandines, is an intrinsic property of their petals then what is the status of the fields of ultraviolet visual quality possibly seen by bees in the central region of the flowers? They must be elements of mental images within the bees' visual systems. (Or, alternatively, if the bees see their colour directly then the uniform yellow apparent to man must be an artefact of the human mind.)

The issue of sensory quality conflicts arising out of animal and human perception poses one of the greatest possible challenges to direct realism. One imagines that it should be impossible for both a person and, say, a bee to simultaneously see the same part of the world directly and yet for both of them not to experience it as having the same colour appearance. Worse still, the problem extends far wider than just the sense of vision that we have concentrated on so far here. Sheepdogs are famously able to respond to the high-pitched whistles of their keepers which humans cannot hear and bloodhounds with their highly-developed sense of smell can follow odour trails which many other animals, let alone humans, are unable to detect. Do qualities of sound and smell exist in the environment in such cases even though silence and absence of smell are all that are apparent to man? Then there are senses such as magnetism and electric-field detection possessed by certain animals that are in their entirety completely beyond the experience of man. Are the external, physical, information-bearing signals on which they depend correlated with sensory qualities in the same way that light and sound waves are, even though they are entirely inaccessible to the sensory consciousness of humans? If our new form of direct realism is going to be a viable theory it is vital that it should offer a means of overcoming the seemingly insurmountable difficulties which all such cases present.

THE POSSIBILITY OF MULTIPLICITY

The fundamental problem which human–animal sensory conflicts raise is that they seem to require that parts of the physical world possess more than one instance of a sensory quality at the same time, at least while under observation by differently equipped species of animal and human. Thus – concentrating for the moment just on the sense of vision – in order to account for all of the visual experiences that may occur during simultaneous observation by humans and bees, individual plants such as grass and flowers each need to be able to possess a number of colours at one time. The grass, perhaps both red and green, and flowers like the lesser celandine, yellow together with an 'ultraviolet' quality known only to the apian world.

This suggestion of a simultaneous multiplicity of colours on objects seems impossible, even (some might claim) ludicrous. However, our relational theory of sensory qualities makes it intelligible how such a multiplicity could occur. We can see this initially just by thinking about the nature of relations. Consider, as an illustration, the relationships of position that are entered into between books stored in neighbouring rows of free-standing bookshelves (of the type which are often found in libraries). Such an arrangement makes it possible to see in abstract how a single object may possess numerous relationships at one time. Take, for example, one of the books at the front of a high shelf. It could be at a relative position of '30 cms above' another book on a low shelf, but at the same time also '20 cms to the left of' a neighbouring book on the same shelf, and '10 cms in front of' a book in an opposite row of shelves. The book is engaged simultaneously in many relationships of position with all of the others on the remaining shelves. In exactly the same way, under the new relational conception of sensory qualities the possibility exists that an object could present as relationships multiple colours simultaneously to various points in its surroundings, or even to different observers. Hence a blade of grass could in principle have one colour-presenting relationship with an animal observing it (such as a bee), in which it presented, say, the hue of red, and at precisely the same moment a different one with a human observer, in which it presented, say, the hue of green.

KITCHEN OF MANY COLOURS

In order to examine this further, let us return briefly to our conceptual laboratory of the uninhabited kitchen. Recall that this is a well-lit room containing a wooden table on which stands a vase of daffodils. But this time rather than setting up just a single pane of coloured glass next to the vase we shall surround it with three such panes, each of a different colour. As before, they will be mounted vertically and set up on the table around the flowers. There will be, say, a red pane, a blue one and an orange one. According to our new theory, as a result of this intervention, the daffodils are made to present four different colours simultaneously to distinct regions of the room. This comes about as follows. The flowers will present green beyond the blue pane, just as they did before when we used a single pane of this colour. But orange will be presented by the flowers to the region beyond the red pane because it filters out all but wavelengths equivalent to 'orange–light' from the range reflected from the 'yellow' daffodils. Because only wavelengths equivalent to 'brown-light' pass through an orange pane from the flowers, the colour brown will be presented beyond it. Finally, yellow continues to be presented to all regions of the kitchen which reflected light from the daffodils can reach without passing through any of the coloured filters.

This is the wonder of relationships, and relational colours, at work. As we noted in the case of the library books, at a single moment one object can be in numerous relationships with others. Or, as here, an object, such as a daffodil, can be in multiple relationships with different regions of its surrounding space. Here those differences in relationship are caused by the distinct wavelengths of reflected light which pass into the different regions. To each of those regions a daffodil presents distinct colours simultaneously.

Returning to the issue of animal vision, we can now see that the fact that objects are seen to have different colours by human observers than by animal ones, is not in principle an insurmountable difficulty for the relational account. What is required in order to fully explain the multiplicity of apparent colours is some notion that when an animal sees an object such as a flower, it participates in the physical relationships that the flower has with its surroundings. The same goes for any human who sees it. And that in both cases the observers participate in such a way that they determine the colour which the flower presents to the point which they occupy, in a similar manner to that in which the information content of light does to points which the light passes through. If, finally, the animal can determine a different colour to be presented than the human (analogous in the example of the daffodils in the kitchen to the former taking the role of, say, a blue filter and the latter an orange one) then we have an explanation in principle of how the flower can simultaneously present different colours to humans and to bees. An explanation, furthermore, which does not require that the perception of either the human or the bee is mediated by a representation laden with sensory qualities, and which permits that it can be direct.

RELATIONSHIPS TO BRAINS

To produce such an explanation let us go back to the basics of our account of how sensory qualities are presented. Recall that information in the signal passing from a 'source' to a location in the surrounding space qualifies the relation between the two. In doing so, it determines the sensory quality which is presented in a relationship from the former to the latter.

This occurs whether or not anyone is present. But now what happens if an observer – human or animal – enters the scene? Information is an abstract structure that is not tied to any specific form of physical embodiment. So when the signal, whether it be light, sound waves or odour molecules, impinges on the observer's sensory organs it is converted by a process known as 'transduction' into electrical impulses which are conveyed along nerves into the brain. These impulses maintain the information that was originally incorporated in the signal, but express it in the somewhat different form of patterns of neural excitation.

(Evidence suggests that this is so because nerve cells in the sensory pathways of the brain have been detected as firing in response to the information in sensory stimuli that is associated with the experiences of colour, sound, smell and so forth.) The information is conveyed onwards in the form of patterns of impulses throughout the network of neurons which makes up an observer's sensory system.

As a result, the sensory system (either in its entirety, or perhaps a process within it which lays down memories of the impulse patterns[6]) can be envisaged as enjoying a similar type of information-determined relationship with the source of the stimulus signal as those points in the space surrounding the source through which the signal passes. It is the recipient of the same type of information as them. So just as the relationship of a location in space with the source is qualified by information carried in the signal, so the sensory system's relation with the source will also be qualified by that same information. But this means that if the sensory qualities presented from a source location to the points surrounding it in space are determined by the information borne in the signals that pass through those points, there must also be a determination of the sensory quality presented to any sensory system into which the information in the signal is imported by transduction.

HOW MANY COLOURS?

To take an example, consider a lime in a bowl of fruits. This reflects ambient sunlight at 550 nanometres ('green' light) to the surrounding space and so relationships of 'presenting' are set up between its surface and each of the points the light passes through (as a result of which the colour green is determined as being presented). If a human looks at the lime, the light entering his or her eyes is converted by transduction into electrical signals by the photosensitive pigments in the rod and cone cells of the retinas of his or her eyes. The information about the fruit's microphysical surface details that was previously expressed in characteristics of the light such as its wavelength, is now expressed in features of the wave of neural excitation which travels up the optic nerve and along the pathways of the viewer's visual system. This means that a relationship of presenting is set up between the lime's surface and the observer's visual system (either as a whole or to its primary memory storage process) in which (as with the point of empty space in the observer's absence) the colour green is determined as being presented.

This does not mean, however, that when an observer looks at a coloured object such as the lime, two overlapping versions of its colour quality exist: one version presented to the space which a person currently occupies (being the version which was presented to the point in space before his or her arrival there), and the other version presented to a location somewhere deep inside the observer's brain which physically resides at the same place in the universe. The point here is that what occurs at the 'destination' end of a presenting relationship (as we saw in Chapter 10) is an abstract event of access to the field of colour quality on the source object. It is not a replication of that colour as a miniature image or suchlike.

So what we have in the case of the lime is the colour green located as a field of sensory quality on the fruit's surface, and the occurrence of access to that colour at vast numbers of points throughout space (all those through which light reflected from the fruit's skin passes). These points include one immediately before the observer's eyes, and others even within the interior of his or her eyeballs. There is also a further occurrence of 'access' to the surface green that occurs within his or her brain. However, the field of green colour quality itself only ever occurs on the skin of the lime.

The Distinctions of Species

Now we can proceed with the challenge of providing a detailed account for conflicting human–animal perception. Biologically different types of observers have differently constituted sensory organs. As a result, different sub-sets of information in the signal stream coming from source locations pass into their perceptual systems. We have seen that bees, for example, can pick up information expressed in ultraviolet wavelengths of light arriving at their eyes whereas humans cannot. Furthermore, perceptual systems differ greatly from one type of animal to another in terms of the internal arrangement and interconnection of their neurons. This means that as the electrical impulses get processed through the perceptual system of each species, distinctions are introduced into the information which the patterns of impulses embody. Because they stem from its neural architecture, such modifications are inevitably unique to each species.

So once it is deep within a species' sensory system, the information embodied in patterns of neural excitation takes on a character which is unique to that species, and which is derived from the organisation of its central nervous system. The distinctive nature of the information to be found within each species' sensory system qualifies the relation between source objects and the animals' sensory systems differently. As a result, in each case a different sensory quality may be determined in relationships of presenting from source objects to the sensory systems.

To illustrate, let us see this in action with light and vision. Take a blade of grass. Imagine that it is growing in a garden which is simultaneously being tended by a human and flown over by a bee. The blade of grass reflects a mixture of sunlight which consists predominantly of 550 nanometre ('green') light, but also contains some ultraviolet wavelengths. When the bee and the human look at the grass, a first difference between them is that only the bee will pick up information in the light reflected from the blade that is expressed in the ultraviolet wavelengths. But then there are further differences, because the organisation of the bee's central nervous system is very different from that of man's. As a general rule, while information is being processed it changes, even if only to become either more or less complex. The internal organisation of a species' neural structure is bound to have an impact on this process. As a result, as it is processed through the bee's visual system, information from the blade of grass takes on features derived uniquely from apian neural architecture and therefore becomes unlike the information at equivalent stages in the human system. It follows from these dissimilar information contents that the relation between the grass and the visual systems of the human and the bee are qualified differently. A different colour quality is thereby determined as being presented from the blade of grass to the bee than to the human. (Perhaps green to the human and red to the bee, as Downer proposed.)

This account of the difference in presented colour quality is in principle not dissimilar to the situation where it is derived from different information being transported by distinct wavelengths of light to two separate points in space. The only distinction here is that in the present situation the difference in information arises because of contrasts in what goes on in each perceptual system rather than differences in the light itself.

CORTEX AND CONNECTIVITY

A significant implication of this account of animal-human visual conflict is that it means that a difference (or equivalently a change) at the 'brain' end of a relationship of sensory quality presenting can bring about a difference (or change) at the source end. This follows because, according to our theory, it is the internal difference in structure of information within bees' and humans' visual systems that is responsible for the difference of colour quality that is presented to them from the surface of the blade of grass they are perceiving. This suggests the existence of, and probably requires the existence of, a fundamental interconnection between brains and places in the surrounding universe with which they can have the physical relationships that we call sensory quality 'presentings'. For the time being we will just note this as a prerequisite of the theory that is being developed and will reserve thinking about it further until later (in Chapter 17), when we will consider the relation between sensory qualities and space.

INFINITY OF PRESENTINGS

Another important consequence of the account is that objects can present an unlimited number of colour qualities to the various observers around them. Thus in addition to the green which it presents to its surrounding space, a blade of grass may, as we have seen, present another colour such as red to a bee. Yet distinct subsets of the wavelength-mix reflected from the surface of the grass will also be picked up by the many other species of animal that observe it, such as birds which fly above the garden and all of the other types of animal which from time to time pass nearby. Each of these species of animal has a unique neural architecture to its visual system. The grass may therefore present somewhat different colour qualities to each such animal, meaning that the number of colours which it presents is effectively infinite.

This plethora of sensory qualities in nature draws our attention to another significant feature of relationships. Unlike quantitative units, such as the atoms and molecules which make up the world of matter, they are not additive. We have seen that in principle an object can enter into multiple relationships, but it is also true to say that there is no limit to the number of relationships that an object can enter into. This can be seen by returning to the illustrative scenario of books arranged in parallel sets of free-standing shelves. If a book was placed on the shelves while empty, and then they were progressively stocked, the original book would gradually acquire more and more positional relationships with the others. But there is no sense in which the first book's acquisition of a growing number of relationships would make it increasingly difficult for it to enter into further new ones. (Librarians do not find it progressively more difficult to stock shelves as they fill up.) There are therefore no limits to the number of relationships that an entity can enter into.

'PRESENTING-TO'

But doesn't the idea that objects have multiplicities of colour raise a different problem? For it seems to require that their surfaces are 'crammed' with fields of colour quality. Take the blade of grass, for example, which may present red as well as green and numerous other colours to the nearby environment. In Chapter 10 we stated that this amounted to points in the surrounding space having 'access' to the fields of colour (red, green and so on) of visual quality which themselves were located on the object. Such a description depicts the arrangement in terms of the colour qualities being spread out across the surface of the blade of grass, and points in surrounding space (by virtue of the light passing through them) acquiring access to those fields. But now, confronted by the issue of multiplicities of colour, we can see that this depiction – although it has served our purposes so far – is somewhat too simplistic. For if we are to take on board the explanation just provided for conflicts between animal and human colour vision, it would carry with it the implication that there are large numbers of conflicting fields of visual quality packed onto the surface of every object. This surely cannot be true. Besides which, the claim that colour qualities are embedded in the physical surfaces of objects is overly suggestive of the paradigm of exhaustive containment.

What is called for are new ways of characterising the elements of colour which exist on the surfaces of objects, and those which allow 'access' at points in the surrounding space. I think that what we have to do – as so many times before – is to recast our thinking in terms of relations. (This will call for the use of some rather unnatural and technical sounding terminology, but I see no way of avoiding this.)

Given that we know that colours are relations, to think of a field of colour quality on a surface as *one* entity and a point in surrounding space which accesses it as *another* separate entity is rather like, in the relationship of object A being 'next to' object

B looking for an element of 'nextness' somewhere in the physical make up of A and another element of 'nextness' within the physical make up of B. Relationships simply are not like this. Just as they are not additive, so they are not composite. They are not made up of constituent subsidiary elements. Rather, they are whole or they do not exist at all.

Let us apply this line of thought to the issue of colours and their surrounding access points in space. Knowing as we do that colours are relations, according to the holistic nature of relationships, in truth we should say that the field of visual quality and its surrounding access points are simply opposite ends of *one* entity – the physical relationship which is colour. If one was forced to characterise the field-of-visual-quality 'end' of the relationship in isolation, acknowledging its relational dependence on the points of access which surround it, one might attempt to do so as a colour 'quality-presented-to' these access points.

This would involve saying that the surface of the grass, for example, contains 'greenness-presented-to' the points through which green-light passes and 'redness-presented-to' the points occupied by bees' visual systems which process information from the grass in their distinctive way. But it contains no intrinsic greenness and no intrinsic redness independent of the points of access around it. Given the holistic nature of relationships it is doubtful whether such descriptions can ever amount to anything more than a partial account of them. Nevertheless, they help in clarifying how the holistic nature of colour relations permits our account to avoid the apparent difficulty of multiple colour fields packed on a single surface.

For on the blade of grass its surface is not required by our account of animal visual experience to contain two conflicting coatings of redness and greenness that are embedded in its structure as such fields of colour quality are traditionally conceived under the paradigm of exhaustive containment. On the contrary, the redness and greenness are each an end of a holistic physical relationship across space *to* either the points in the local environment through which light reflected from the grass passes, or viewers in the environment who process the information which the light carries. There is no problem in principle about an object's surface being the site of numerous ends of different such relationships. In other words in principle it could have not just two but many – possibly even an infinite number – of such colour 'quality-presenting-to's.

THE SHEPHERD, THE SHEEPDOG AND THE ULTRASOUND

The broad principles we have used in this chapter to explain how objects can present multiple colours simultaneously to different classes of viewer can be applied to modes of sensing other than vision. In this form they are able to account for most and perhaps all cases of animal–human sensory conflict. To illustrate, we will consider next an example case based on the sense of hearing. Imagine an event which generates sound waves that are detectable by humans, but also generates ultrasound at a frequency beyond man's capabilities. Let us say it is the squeal of rubber against tarmac produced as a car brakes sharply on a country lane.

The sound waves reach the ears of both a sheep dog and those of the shepherd in a nearby field. Just as with colour, it is the information content of the sound waves (here, concerning the thousands of microscopic vibrating impacts made by the tyres on the tarmac each second due to friction) that determine presented sound at the points in space which the sound waves pass through. But this information also passes into the shepherd's and the dog's auditory systems, when the sound waves are transduced by auditory hair cells in their inner ears into electrical impulses, which travel up the auditory nerves into their brains. So, on similar grounds as with vision, it follows that these auditory observers participate in information-based relationships with the sound wave-generating event (of the tyre-rubber sliding on the road), in such a way as to be able to determine the sound quality presented from it to each of themselves. But then it also follows that any differences in organisation or structure of their auditory systems will lead to corresponding differences of presented sound qualities.

Here we have precisely this kind of variation, because only the sheep dog's ear is able to respond to the highest frequency of sound waves emanating from the screeching tyres. The dog's auditory system is able to absorb the full spectrum of information available in the sound waves and to convert it into patterns of neural excitation, but the somewhat paltry human ear of the shepherd detects only the lower and mid-range frequencies available and hence absorbs some but not all of the information content. As a consequence, the sheep dog's auditory system determines a different sound quality to be presented to it from the event than does the shepherd's. However, as in the case of the bee and the human seeing a blade of grass with incompatible colours, such sound qualities, despite being different, are not in the minds of the observers. The presented sound is a relationship – in this case from the event in the external world to the observer. It is this relationship which has been qualified by the distinctive forms of information-uptake available within each species' auditory system.

Spectrum of Qualities

What about conflicts arising from sensory qualities which may conceivably be associated with those senses operating in the animal kingdom that are not possessed by man? For example, certain animals can detect the infra-red radiation that is given off by warm bodies, others (such as electric eels) electric fields, and then there are the 'echo-locators' like bats and dolphins which undoubtedly acquire a very different aural picture of the world than man. It seems plausible to think that relational sensory qualities may exist in the physical realm in association with the information-bearing signals that are utilised by each of these senses. There is also the question of the existence of sensory qualities which may potentially be associated with information-bearing signals that as far as we know have not yet been exploited by the sensory organs of the animal kingdom (such as gravity).

We cannot be certain whether any of these information-bearing signals determine sensory qualities in the physical realm, but equally, consistency would seem to demand that we should allow for the possibility that they do. That being the case, there may be many more situations than those we have considered so far in which an object appears to humans to have one quality but objectively presents many others. For example, imagine a copper wire which is heated by the passage of an electric current, to the point where it glows red. This presents sensory qualities that are detectable by humans of heat and colour, but also, because the electric current creates an electric field around the wire, possibly also a quality associated with the field, of a kind that humans cannot access (but in an aquatic environment electric eels would). In a parallel way many living organisms are also coloured, warm and surrounded by weak electric fields (due to the electro-chemical nature of their cells) and so may also present a similar array of sensory qualities to the environment. Then, of course, there may be a whole spectrum of such supra-human sensory qualities occurring objectively in nature as a whole that we are entirely unaware of. From qualities possibly associated with magnetism, the ultraviolet, the infra-red and ultrasound to conceivably even gravity. Can the relational direct realist approach accept the idea in principle of such a multitude of sensory quality types within the physical realm that man has no consciousness of? Although undoubtedly counter-intuitive there is no logical difficulty about this, given the theoretical framework that we have provided ourselves with.

The difficulty presented by objects possessing multiple supra-human sensory qualities is in logical terms simply an extension of the problem posed by them having many colours simultaneously, and so is addressed by the same solution. What we are asking of objects is that they should have at one moment not just, say, the single colour visible to us on them but also an almost infinitely large number of other qualities, each one associated with an information-bearing signal (electromagnetic field, heat-radiation, etc.) emanating from the object into surrounding space.

As we saw earlier with colour, this does not require that the object possesses untold numbers of fields of quality embedded in its surface. Rather it is to require only that the object be engaged in multiple presenting relationships based on each of the types of signal. So the existence of numerous sensory qualities beyond those known to man does not mean that there are multiple fields of sensory quality 'crowded' onto the surface of each object. The exterior of the heated copper wire, for example, does not have a coating of colour quality, another of heat, together with a further one of 'electric field' quality. Rather points in surrounding space, through which light from the glowing wire pass, have access to its quality of 'redness-presented-to' directed at them, those through which radiant heat energy pass have access to its quality of 'heat-presented-to' directed at them and a similar 'appearance-presented-to' based account can be given for the electric field quality. Also, as we noted earlier, relations are not numerically cumulative so no problem in principle is presented by the notion of objects being engaged in vast numbers of them.

If this is correct then an enormous spectrum of relational sensory qualities *may* exist objectively in the physical realm. Those that man is aware of, while objective and directly accessed by him, would represent merely the tip of the iceberg of all of those which fill the universe. (I do not claim that such a spectrum definitely exists. Only that the relational form of direct realism is compatible with the possibility of its existence.[7])

LOCKE'S BUCKET REVISITED

Having extended the relational account of sensory qualities to encompass the problem of animal-human sensory conflicts we are now in a position to address one of the more mystifying arguments against direct realism. This is the point put forward by John Locke that the temperature felt when one immerses one's hands in tepid water depends on whether they start off as hot or cold. A cool hand will feel the water as hot whereas a warm hand will feel it as cold. There is no doubt that this phenomenon occurs (you may like to try it at home) and on the face of it, it seems to suggest that the sensory quality of heat (or cold) cannot be an inherent property of material bodies. As Locke says, 'it is impossible, that the same water...should at the same time be both hot and cold.'[8]

However, we know otherwise. It is not in fact impossible in principle for bodies to present conflicting qualities to their surroundings and for those qualities to be objective elements of the physical realm, when they are relationships. Furthermore we have also established that the visual and auditory systems, and by extension sensory systems in general, can play a role in determining the sensory qualities presented to themselves (as a consequence of differences in the way that they take up or process sensory information). This provides an adequate basis to account for the 'conflicting water temperature' phenomenon noted by Locke. Prior to their immersion, each hand is in a different state with regard to temperature. This means that when they are placed in the same warm water the liquid will enter into *different* relationships with *each* hand.

For example, there will be a relationship with the cold hand which in physical terms involves a large flow of heat energy from the water to the hand. In contrast, the relationship with the hot hand involves only a comparatively minor flow of heat energy – indeed if the hand is hotter than the water the energy transfer will be reversed. Just as for the qualities of colour and sound, the sensory quality of heat presented from an object to a point in space is determined by the information borne in the energy flowing from the object to that point. With the quality of heat this information is expressed simply in the quantity of radiant or convective heat energy flowing to the 'destination' point. Thus the greater the energy flow radiated from an electric fire (the higher it is turned up) to a point in the surrounding space, the hotter the temperature a person is able to sense at that point.

The relationship between an energy emitting source and the heat sensing 'organ' of the skin which determines the presented quality of heat for a perceiver, is qualified by any respects in which the information is captured or processed differently by the skin – just as colour can be determined differentially by the processing characteristics of the visual system. What we see in the case of 'Locke's bucket' is that the skin of each hand absorbs different quantities of convective heat energy and thus different heat-pertaining information from the surrounding liquid. The information which the hot hand picks up is equivalent to that which a hand at room-temperature would pick up from cold water, because the quantity of in-flowing energy is low. This information transfer determines a quality of coldness to be presented from the water. At the same time the cool hand receives a large inflow of energy and thus of heat-pertaining information. This information transfer determines a quality of hotness to be presented from the water. Locke's paradox is overcome because these are 'relational presentings' of heat and cold from the water which may – contra Locke – occur simultaneously from the same body of water. (It is no more contradictory that the liquid should have two of them than that a book in a library should have multiple relationships with others in the surrounding shelves or that a blade of grass should present green to a human and simultaneously red to a bee). Accessibility to the sensory quality of heat only arises in the cool hand because of its informational state. While, simultaneously, accessibility to the quality of coldness can only arise in a sensory system which is in the informational state that occurs in the hot hand.

16. Jewel in the Sky

The next problem to be considered is that raised by illusions. There is no question that these do not ordinarily represent real physical objects. When the sufferer of a drug-induced hallucination 'sees' green 'blood' oozing between paving slabs there is no physically real fluid between the paving slabs (only at best perhaps some meagre outcrops of moss). But there remains the possibility for the direct realist that the sensory qualities perceived in illusions are real and external. So even though there may be no abnormally-coloured blood underlying what is seen in such a case it is still conceivable that the person undergoing the illusion may be experiencing a green colour quality which is itself genuinely located throughout the external region which the oozing liquid appears to occupy (i.e. filling the gap between the slabs). This is all the more possible now that we are pursuing a relational conception of colour. Thus the fact that nobody else sees the external green could merely be because they are not engaged in the same relationship with its location. It is not necessarily indicative of the non-existence of the sensory quality at that location in the external realm.

INFORMATION AND PERSISTENCE

The easiest place to start developing an account of illusions is by looking at micro-illusions such as after-images and similar 'after-effects'. Let us imagine that a person is gazing into a powerful table lamp. In the previous chapter we worked out that the wavelength information contained in light flowing outwards from this sort of source transmutes into a neurological form when it crosses the barrier of the retina, and is then converted into electrical impulses travelling up the optic nerve into the observer's visual system. This in turn qualifies the relation between the lamp and the visual system, thereby determining the colour quality presented from the former to the latter.

After-images occur when one looks away from such a light-source, or the source of light is otherwise interrupted, so let us take the case initially that the lamp is simply switched off. This has the effect that light ceases to stimulate the observer's retinas. But it does not mean that information from the light also instantaneously stops flowing within the visual system. In fact, after-images are caused by the tendency of retinal cells,[1] and perhaps others in the visual system, to adapt to the intensive rate of firing brought about by powerful sources of light. When the source of stimulation ceases (in this case the lamp is switched off) the firing continues for some moments until it eventually subsides and in consequence the after-image fades. But because such 'adaptive' firing of cells occurs in the same pattern as the original excitation, during this period *identical information* is sent into the visual system as when light from the lamp itself was stimulating the eyes. Therefore while adaptive firing continues, the relation between the lamp and the viewer's visual system remains qualified in the same way as it did when the information in the viewer's brain was derived from the flow of the lamp's light into his or her eyes. The effect of the existence of an identical relation between the lamp and the observer's visual system as when it arose from light entering his or her eyes is to bring about the determination of the same colour quality in the relation between the two. This means that the colour quality that was presented by the lamp to the observer during vision when mediated by light, continues to be presented while the same information is maintained in the visual system through adaptive firing. The colour of the lamp in this case is (bright) yellow, so the viewer sees a continuing 'blob' of yellowness presented from the location of the physical light-source after it is switched off. This lasts as long as the adaptive firing of cells continues within his or her visual system and the original information from the light is maintained there.

BLOBS IN SPACE

Now let us consider the case where the after-image arises not as a result of a lamp being switched off, but because the observer turns his or her gaze away from it. Here the adaptive firing of cells in the visual system maintains information from the lamp as before, but the relationship which is qualified as a consequence cannot be with the physical matter of the lamp itself, because it is no longer within the observer's field of view. (In more extreme cases, it may not even be in the vicinity if, for example, he or she has been spirited away in a high-speed vehicle.) Instead, the relationship is with whatever region of space is, given the turning movement of the observer's head, currently in a position relative to his or her brain, and visual system equivalent to that which the lamp occupied when the adaptive firing was taking place.

This position is normally a region of empty space a short distance in front of the observer's eyes. Hence the characteristic way in which the 'blob' of yellow seen as an after-image appears to hang in space before one's eyes and moves with one's direction of vision as one looks around.

What all of this suggests is that the colours accessible in an after-image arise in essentially the same way (and in the same location relative to the observer's brain) as when a person sees an actual lamp. The principal difference is that the after-image case depends on the persistence of information inside the visual system. Thus in seeing a physical lamp, its presented colour is made accessible within the visual system as a result of information flowing from the lamp firstly in the form of light and then, once converted by transduction at the retina, as electrical signals. In experiencing an after-image, the presented colour is made accessible within the visual system as a result of the persistence of the identical information.

In the previous chapter, our account of animal sensory consciousness indicated there must be a connection between the brain and the surrounding space such that a difference in an observer's brain can have a determining impact on the sensory quality presented from a 'source' location. The present case offers an example of this.

Thus when the adaptive firing of neural cells gives rise to after-images, the information that flows into the brain determines that a point in space in front of the observer (equivalent relative to him or her to the original position of the lamp) presents the lamp's original colour to the visual system. This colour (say, yellow) is determined as being presented from a region of space of the observer for as long as the information persists in his or her visual system due to the on-going firing. So it is that a perceiver experiences a 'blob' of yellow apparently hanging in the air in whatever direction he or she looks. The colour quality which makes up this illusion is an external one that is presented from, in this case, a region of space immediately in front of the observer. Other observers do not experience the same external colour because their visual systems do not contain identical information, as they haven't stared at the lamp, and hence the location in space is not in the same relation with them.

Micro-illusions such as after-images can therefore be explained without having to account for them as elements of a visual representation mirroring the external world. They are relational presentations of colour from regions of space ahead of the observer, brought about by the persistence of information in the visual system.

MICRO-ILLUSIONS AND MEGA-ILLUSIONS

Having established a way of accounting for micro-illusions which is compatible with the direct realist outlook of our theory, let us now attempt to apply it to more severe cases of illusion.

We will start with drug-induced visual hallucination, the most dramatic form of illusion of all. In fact, even under the influence of hallucinogenic drugs it is rare for an image to be experienced which bears absolutely no relation to underlying physical reality. For in most cases, visual illusions, including hallucinations, derive substantially from misinterpretation of what is in view, as in the famous examples quoted by classical Indian philosophers where a rope is visually mistaken for a snake or a shell for a piece of silver. In contrast, those supposed cases sometimes cited by Western thinkers, where people under the influence of alcohol are said to see images such as 'pink elephants' that are entirely unrelated to the visual background which they overlay, seem to have little basis in fact. (They appear to owe more to the naïve, pictorial depictions of illusion that are to be found in comic-books, in which characters see 'stars' literally in front of their eyes, than to any genuine human experience.) Nevertheless, in order to work with the worst possible case for direct realism let us imagine that just such an 'image-type' hallucination has arisen in the field of view of a person who has ingested a tab of LSD. To make things more specific we will say that when this person looks at the world around himself, what he sees in the centre of his field of vision is a large symmetrical green jewel. Just like an after-image it overlays objects in view, and its apparent size depends on the scale of what is seen around it, so that when the person gazes up in the sky he sees it as of enormous size and apparently floating high above the clouds.

As in the case of an after-image, this illusion is not the result of light currently entering the LSD-taker's eyes and conveying optical information from a giant, physical jewel-stone in the sky. The explanation of after-images lay in the fact that the relevant optical information persisted and continued to be processed in the observer's visual system. So, could something comparable be going on here?

In order to obtain an answer it is necessary to appreciate how drugs such as LSD operate at a neuro-chemical level. During the normal waking state of daytime, certain circuits of neurons in the brain secrete a substance called serotonin which inhibits the neural mechanisms that underlie the process of dreaming[2] and thus prevent dream imagery from intruding on day-to-day experience. Drugs like LSD and psilocybin which are capable of producing vivid visual hallucinations similar to the 'green jewel' have the effect of suppressing the activity of these neuron circuits.[3] As a result, the drugs in effect enable the mechanisms of dreaming to reassert themselves in a conscious context, and thus permit images such as one might normally see in a dream to be encountered in the waking state. So to understand further what is going on in the process of drug-induced hallucination one must inquire into how the visual imagery of dreaming arises.

SEEING AND SLEEPING

It is well established that dreaming occurs in the REM (Rapid Eye Movement) phase of sleep. During this form of sleep part of the brainstem known as the 'pons' activates elements of the dreamer's visual system by setting in motion fast eye movements and generating activity in his or her visual cortex. A number of experiments have shown that the eye movements which a person makes while dreaming correspond fairly well with the content of the dream.[4] This indicates that the pattern of neuronal firing in the visual cortex during a dream is similar to that which would have occurred had the events of the dream actually played themselves out for real before the dreamer's eyes. In other words, even though it has been internally generated, the same optical information is being processed by the person's visual system as would have been the case if the events had been real.

So if LSD and similar drugs activate the mechanisms of dream imaging during hallucinatory episodes, what this suggests is that the illusions which they produce are generated in a similar way. In other words, they stem from activity within the neurons of the visual cortex which has been self-generated, as a result of the chemical behaviour of the drug, and which corresponds to that which would have occurred had the object being hallucinated been physically present in view. Or, putting that in terms of information, the chemical activity of the drug on the neurons of the visual cortex brings about the generation of optical information there, equivalent to the information normally produced by seeing real objects. But we have already seen that the existence of optical information in an observer's visual system, due to persistence of neuronal firing, results in the determination of colours that are presented to him or her from a region of space in front of the head, and which take the form of after-images.

There is therefore every reason to think that the information resulting from the action of a hallucinogenic drug should do the same. Despite being self-generated by the visual system, the resulting information would determine colours to be presented from points of space in the field of view ahead of the observer – those points which when occupied under normal circumstances by the object apparent in the hallucination generate the equivalent information. So the determination of a set of colours occurs, as being presented from a field of points in the space ahead of the drug-taking observer. Together these colours form a hallucinatory image apparently hanging in space in front of him or her.

SELF-GENERATED INFORMATION
So this is the explanation of the green jewel (and whether or not hallucinations such as it ever genuinely occur it is certainly the case that consumers of LSD experience 'image-type' illusions

equivalent to the green jewel with their eyes shut). It arises because the person's visual system self-generates within itself – through the mechanisms of dream imagery which are liberated by the drug – the pattern of optical information for such an object. As we already know, the processing of optical information determines the colour which may be presented to a perceiver from a location. In this case, because the information happens to correspond to that for a green jewel the colour green arranged in the shape of a jewel is presented to the subject from wherever in space ahead he or she looks (the hallucination, like an after-image moves with the field of vision as the head is rotated). It is not presented to anyone else because, not having consumed LSD, their visual systems do not self-generate the identical internal optical information. The same explanation can also be given for the illusory green 'blood' which we considered at the opening of this chapter. Here LSD has brought about optical information for such a coloured liquid in the drug-taker's visual cortex (the process having perhaps been sparked by the existence of genuine patches of green moss between the paving slabs). As a result, the colour green is determined as being presented to the drug-taker from the regions which under normal circumstances would produce the same information – in other words the volumes of space between the slabs that would be occupied by a viscous liquid oozing between them. Illusory green is presented from these regions (filling in between the patches of 'genuine' green presented from the outcrops of moss).

It becomes evident that visual illusions, after-images and drug-induced hallucinations can all be accounted for in surprisingly realist terms, with the assistance of the relational conception of colours. There is no material entity corresponding to any of these images but the colour qualities which comprise what is seen in such illusions are located in the external world. In each case, they result from the unusual circumstances in which the brain processes its optical information, such as that it persists as a result of adaptive firing or is produced through self-generation by the action of drugs. But whatever the individual reasons, the results are a pattern of optical information in the visual system which corresponds to that which would have occurred had an equivalent object to the illusory one actually been seen. This pattern determines colours to be presented from locations within the visual scope ahead of the subject, and so produces the illusory image.

THE WEIRD WORLD OF HALLUCINATION

Illusions and hallucinations occur in senses other than just vision. For example, we noted before that the 'ringing' sensation which can be heard after a loud rock concert is the equivalent in the sense of hearing of an after-image in vision. Also, prolonged use of cocaine can lead to auditory hallucinations[5] and heavy users of that drug are known to develop tactile hallucinations such as the feeling that small insects have burrowed under their skin.[6]

But the form of explanation which has been offered in the case of vision can be used to account for all such illusions across the full range of man's senses. In each case, unusual information processing occurs within the brain and sensory system of the subject of the illusion. Under normal circumstances, the information in question permits the subject's system to participate in determining a sensory quality – sound, smell, taste, whatever – that is presented from objects (in the same way that information derived from light permits observers to participate in determining colour). The processing of this information, under the conditions which give rise to the illusion, makes this same sensory quality be presented to the perceiver not from an object, but rather from a location in the external world which is within his or her sensory scope.

Consider cocaine-induced aural hallucinations. The sensory quality of sound is normally determined by auditory information that is conveyed by sound waves. But in the act of hearing this auditory information is further transmitted from the inner ear by nerve signals along the auditory nerve to the brain. By analogous arguments to those which were used in the case of vision, it is possible to conclude that if any self-generated auditory information occurred within a person's brain this would – even in the absence of transmitted sound waves – lead to the presentation of sound from a location within the auditory scope of that person. Something of that nature must be precisely what happens after extended use of cocaine. The drug induces a pattern of activation within the auditory parts of the cortex equivalent to that which would have occurred if a real sound had been heard. As a result (hallucinatory) sound is presented to the cocaine-taker from a point external to him or her.

This last of the major arguments against direct realism based on 'internal' factors, the argument from illusion, has been one of the toughest nuts to crack. But once again the new relational theory of sensory qualities has shown itself able to meet the challenge. Illusions are not the result, as the prevailing picture of perception would have us think, of the perceiver's internal mental representation of the outside world generating a false experience. Rather the sensory qualities which constitute them are located externally to the perceiver and are the product of various forms of anomalous information processing within his or her sensory systems. There is no necessity to invoke a mediating quality-laden representation and we can simply conclude that these external colours and other qualities are perceived directly.

17. The Colours of Space

Next we come to the arguments derived from our current body of scientific knowledge. As we saw in Chapter 11 there are essentially two of these, both associated with space. The first concerns the fact that atoms, and so matter in general, have been found by modern physics to consist predominantly by volume of space. Hence there is remarkably little of a substantial nature in material bodies on which qualities such as colour can be 'anchored'. The second derives from the rotation of the Milky Way galaxy and the proneness of stars to explode as supernovae. When combined with the speed of light in a vacuum these latter facts mean that the positions in the night-sky at which the white points of many stars are seen (the positions from where their whiteness is presented according to our theory) consist solely of empty space.

SPACE MAN

The need for us to take the issue of space and its connection with sensory qualities seriously is also strongly suggested by the solutions so far offered for the arguments from illusion and from refraction. Taking each in turn, consider first the argument from illusion. This envisages that the arrangement of sensory qualities which make up illusions, such as after-images and drug-induced hallucinations like the 'green jewel', are presented to the subject from a region of space ahead of him or her. This carries the implication that such sensory qualities may be presented from locations which consist only of empty space. That would most obviously be the case in the eventuality (unlikely but nevertheless possible in principle) that LSD had been taken by an astronaut and his hallucinatory images were presented from literally the vacuum of outer-space. But even for earth-bound drug-takers, hallucinations are typically presented from regions of air made up of a gas of widely spaced molecules (such as oxygen, nitrogen and CO_2) in a background volume of space. So in both of these situations our explanation of visual hallucination commits us to the concept that colours may be presented to subjects from points of empty space.

Furthermore, this is a problem that is not only confined to colours and the visual sense. It also applies to the account of illusions in other sensory modes. Thus, to take just one example from man's full set of senses, the auditory hallucinations which result from long-term use of cocaine, in so far as they appear to subjects to come from the external world, were also explained as sound qualities which are presented from the surroundings. For the terrestrial drug-consumer these locations, as before, consist only of air molecules in a background of space and again it is not hard to imagine circumstances involving cocaine-taking astronauts in which they would comprise the vacuum of empty space.

RAY OF THE KINKS

The same basic commitment to the idea of sensory qualities being presented from empty space applies also in the case of our solution to the argument concerning the phenomena of refraction. In order to explain such appearances as bent sticks and rows of tiles with apparent kinks in the middle, our theory relies on the idea that colours may be presented from positions that are off-set relative to all or part of physical entities, when the paths of light rays have been bent by refraction. But the positions from which such 'off-set' colours are presented could in principle also be ones occupied only by empty space. This would happen, to take another extraterrestrial example (again extreme but entirely possible in principle), if a spaceship's portholes were fitted with 'wrinkled' glass similar to that used in domestic bathrooms. Then the colours of say, a passing satellite as presented to the interior of the spaceship through its viewing windows would be 'off-set' into the vacuum of space next to the physical structure of the satellite, as a result of the refraction of light rays as they passed through the glass.

HONEY, I LOST THE SUN!

In real life of course it is unlikely that the viewing windows of a spaceship would ever be fitted with anything other than optically perfect glass. Yet the equivalent off-setting of colours into empty space occurs every morning and evening when we, as terrestrial observers on earth, look at the sun rising or setting close to the horizon. This is because the earth's atmosphere is a high density medium in comparison to the void of space, so that sunlight is refracted when it passes through it on its journey to the earth's surface. As a result, light from the sun which enters the atmosphere at a shallow angle is bent downwards towards the thicker layers close to the surface. This affects the apparent position of the sun in such a way that, other than when it is straight overhead, it is always made to seem somewhat higher than its actual position. The difference is most marked at sunrise and sunset, when the sun is low in relation to points of observation on the earth's surface and its light rays are entering the atmosphere at the shallowest of angles. It has been established that when the sun appears to lie on the horizon the physical body of the solar globe is in fact below it by approximately thirty-five minutes of arc (roughly its own diameter).[1] So what is a person truly seeing when he or she observes at sunrise the disc of golden yellowness emerging from the horizon in the distant sky? Such an observer is certainly not having a direct experience of the colour of a giant ball of gaseous matter. This is because the sphere of yellow colour that he or she has awareness of is in the wrong place for that to be possible. The physical sun, the seething ball of hydrogen and helium, is in one place – thirty-five minutes of arc below the horizon – while the apparent globe of golden yellowness is located in space that is empty of all matter.

If one considers the earth, as has sometimes been suggested, to be like a 'spaceship' journeying interminably through the solar system, then the atmosphere can be considered to be the porthole through which all celestial objects are viewed. In that case, when the sun is low in the sky the effect of refraction in the atmosphere means that – as with the porthole of a genuine spaceship fitted with wrinkled glass – its colour is presented from off-set positions which are occupied only by empty space.

ARE SENSORY QUALITIES PROPERTIES OF SPACE?

So the relational theory of sensory qualities that we have developed carries with it on a number of counts the corollary that colours and sounds are properties of space. As all sensory qualities share the same essential nature of being relationships, in effect it thereby also commits us to them all being properties of space. Is it possible that this could be true?

On first consideration the idea seems bizarre. Although in day-to-day thought we don't necessarily think of qualities such as sound and smell as properties of an underlying substance, we are used to thinking of colours, in particular, as features of the matter of the bodies to which they belong. So the notion that colour might be a property of the emptiness of space is especially hard to entertain.

Nevertheless, it is not difficult to see how this could be a viable idea. For example, if it was space rather than matter that was coloured, then the colours apparently 'on' solid bodies such as bricks and buses could arise as the temporary current properties of the volumes of space which the bodies occupied. So the notion is by no means inconceivable. In fact, the idea that sensory qualities might be properties of space has already been contemplated. In the late 1910's the innovative English philosopher Alfred North Whitehead proposed the concept almost in passing, in the context of arguing that space is not a relation between substances but between 'attributes'. He suggested that: 'It is not the substance which is in space, but the attributes. What we find in space are the red of the rose and the smell of the jasmine and the noise of the cannon.'[2]

If it is conceivable, then, that sensory qualities are properties of space, the next question is are they so in fact? Next we will look at a number of arguments which suggest that they are.

ATOMS OF SPACE

Firstly, according to contemporary physics the nature of atomic matter is such that sensory qualities – and in particular colour – if considered to be features of external material bodies would in fact *have* to be properties of space. This is because, as we have previously noted, matter has been shown by physicists to consist predominantly of space. To give an idea of just how much space, according to one recent comparison the scale of the nucleus relative to a whole atom is comparable to that of a pinhead compared to the dome of St Paul's Cathedral in London.[3] But it is also the case that sensory qualities are spatial in nature. This is especially clear with colour which as a field of visual quality occupies *areas* on the surface of objects, and in the case of translucent colours, *volumes* within their interiors. It therefore follows that if they were characteristics of material objects, such qualities would necessarily also have to be attributes of those elements of matter which comprise its spatial volume. These elements, physics tells us, are nothing other than the space between the atoms and the space of which the atoms themselves are predominantly composed.

TIME-LAG REFLECTION

Next, consider the time-lag argument. This is normally cited as offering an incontrovertible case against direct realism. To paraphrase the way that the argument is usually presented: 'It is impossible for colours and other sensory qualities to exist in empty space so those experienced in the circumstances identified by the argument (supernovae, stars in rotating galaxy, etc.) must be in the mind of the perceiver.' But this could be to look at the facts the wrong way round. Time-lag perception could instead be taken as indicative of the *necessity* of sensory qualities being properties of space. When a person sees a white point in the night sky coming from a location where there is only empty space it could be argued not that the whiteness must be in his or her mind but that the only explanation is that the whiteness is a property of space. These are radically different perspectives to take on the same phenomenon, and given the prevailing consensus against direct realism, the conclusion that sensory qualities are properties of space is unlikely to find favour unless we can find other arguments which point in that direction.

Nevertheless, this shows at the very least that the time-lag argument – one of the scientific pillars in the 'case' against direct realism – is rarely presented with complete impartiality. Furthermore, by the end of the present chapter, it will be possible to recognise, I hope, an array of such arguments, leaving little room to conclude anything from the time-lag involved in vision other than the fact that colour is a property of space.

QUANTUM FOAM

As we have seen, the way modern physics envisages the nature of matter dictates that sensory qualities cannot be attributed to it without thereby attributing them to space. This means, ironically, that over the centuries man has unknowingly but quite happily been ascribing colours and other qualities to space rather than any genuine 'stuff'. But equally recent developments in physical theory have also transformed man's understanding of the nature of the space that exists between physical objects. Modern science has shown that apparently 'empty' space is not nothingness, and so it becomes conceivable that space could provide the viable 'substratum' for the sensory qualities that the relational account requires.

One significant insight into the nature of space has been the realisation that the vacuum is subject to Werner Heisenberg's uncertainty principle. The Uncertainty Principle is a key concept of quantum mechanics which states that it is impossible to measure at the same time both the position and speed of a particle. A consequence of this is that even the most perfect vacuum is not utterly devoid of matter. Rather, on the microscopically small scale that is contemplated by quantum mechanics it is 'a sea of continuously appearing and disappearing particles.'[4] By 'borrowing' energy from the vacuum these get created as pairs of 'virtual' matter and antimatter particles of opposite charges which annihilate together almost instantly upon creation.

Apart from the fact that most accept the fleeting existence of such virtual particles within space, there appears to be little consensus amongst contemporary scientists regarding its underlying nature. Most recent efforts to delve into the deeper order of space and spacetime have centred on the imperative to build a quantum theory of gravity. This, it is envisaged, would bring together quantum mechanics (the science of the extremely small) and general relativity (the science of gravity and the very large). However, of two leading examples of such work, both suggest the possibility that empty space may have a degree of structure. The first, conceived by the American physicist John Wheeler, centres on the idea that the field of gravity, and therefore the very fabric of spacetime itself, is subject to quantum uncertainty. According to this view, when considered on scales of approximately 10^{-35} metres (the so-called 'Planck length'[5]) spacetime is revealed to be a seething foam of multiply-connected geometry made up of constantly forming and dissolving wormholes.[6]

QUANTUM LOOPS

The second is due to American physicist Lee Smolin and a number of collaborators. It suggests that on a similar scale the fundamental entities are loops of lines of force in the gravitational field. Space is then held to be a relational construct which arises out of the interlinking of these loops. Because the loops are quantised — that is they only come in discrete values like the energy packets of quantum mechanics — space acquires an 'atomic' structure, in the sense that volume itself becomes quantised. That is, 'there is a smallest possible volume. This volume is minuscule — about 10^{99} of them would fit into a thimble.'[7]

Unlike virtual particles, the existence of which is accepted as having been verified, no experimental equipment yet exists to probe spacetime at the dimensions of the Planck length and so strictly-speaking the existence of both quantum foam and quantum loops remains unproven. Nevertheless it is clear, even if only on the better established grounds of the presence of a flux of virtual particles, that the vacuum is far from being nothingness. To quote physicist Brian Greene, 'There is now little doubt that the intuitive notion of empty space as a static, calm, eventless arena is thoroughly off base. Because of quantum uncertainty, empty space is teeming with quantum activity'.[8] Furthermore, as Greene points out, string theory, like Smolin's 'loop quantum gravity', also hints in the direction of spacetime having 'some kind of atomised structure'.[9] This means that the thrust of two leading instances of contemporary physical theory suggest that spacetime — and in the case of loop quantum gravity, volume and hence space itself – possesses a structure.

This is the important point for our theory. For if space has structure – whatever precise form that eventually turns out to take – then it is just as appropriate an entity to support sensory qualities as matter. Indeed more so, given that sensory qualities cannot be attributed to matter without attributing them to space. While it turns out that there is little 'stuff' to matter, empty space in contrast is not nothingness. There is, as it were, a 'something' even to the vacuum devoid of all matter – whether that be quantum foam, interconnected quantum loops or even the flux of virtual particles – onto which sensory qualities can be thought of as being anchored. In these circumstances it is reasonable and plausible to think of sensory qualities such as colour, sound and smell as being properties of space, as is required by our theory.

ARENA OF RELATIONS

Finally, I suggest that space, rather than matter, is the appropriate medium to support sensory qualities when those sensory qualities are relationships. Space is the physical forum of a number of abstract relations such as length, size and relative position. It is therefore readily conceivable that it should equally be so for the physical relationships that are the sensory qualities.

In fact, it may well be that space and the presence across space found in the relational sensory qualities are one and the same. It is sometimes proposed that it is motion that makes space obvious to man but I suggest that it is in fact presence across space which above all else reveals it to him. For it is because we see colours at a distance and sense sounds and smells remotely that we acquire the concept that there exists a field – to which we give the name 'space' – which fills the universe.

A RELATIONAL THEORY OF SENSORY QUALITIES

We have now arrived at the position where the theory that

has been developed can address all of the arguments that are conventionally raised against the direct realist outlook, including those derived from recent developments in science. It is perhaps a good moment, then, to take stock by summarising the position that has been achieved. Our new theory marries a relational conception of the sensory qualities with the basic outlook of direct realism. As established so far, its high-level tenets can be set out in a small number of statements as follows:

- Sensory qualities present across space. As such they are physical relationships over spans of space.

- Considered as properties, the qualities presented in such relationships are properties of space.

- Sensory quality relationships are non-causal.

- The qualities presented in such relationships are determined by information-based relationships. These are typically the result of the information being conveyed by a carrier signal such as light or sound waves, but may also be due to information processing in a brain/perceptual system.

Much ground has been covered since we set out to develop the ideas that have culminated in our new theory of sensory qualities, so let us remind ourselves now of its status. Early in the book we started out on the path of developing a new direct realist account of perception. This was because it emerged (in Chapter 6 and 7) that the concept of micro-mind derived from the direct realist view meant that it had the potential to overcome the challenge of the so-called 'hard problem' of consciousness (the problem of explaining how conscious experience arises from the grey matter of the brain). In this it contrasted radically with the prevailing representational account of perception inspired by Newton, which entailed a 'mega' concept of mind and hence an insoluble hard problem. But the direct realist standpoint itself was confronted by a range of counter-arguments (illusions, refractions etc) and so the strategy we decided to follow was to explore whether a relational account of the sensory qualities might have the potential to overcome these arguments.

That has proved to be the case. In our new theory, the combination of an understanding of sensory qualities as physical relationships (and properties of space) with direct realism has met all of the counter-arguments conventionally posed against the idea that man's sensory awareness is of qualities embedded in the external physical world.

But not only that. During the course of our exploration we have also uncovered two important points. Firstly, we have found that there is widespread evidence in nature that the sensory qualities are relational. Secondly, we have found that any direct realist conception of the sensory qualities as exhaustively contained in their supporting objects – i.e. as non-relational – cannot account for the fact that man experiences such qualities at a distance. Thus it is *only* our new theory's conception of the sensory qualities that conforms to the way that sensory qualities are found in the world. It can also be claimed for our new theory at this point that at the very least it has demonstrated that the direct realist viewpoint on perception cannot simply be dismissed without first disproving that sensory qualities are both relationships and properties of space.

Beyond that, it has also been demonstrated (in Chapter 6) that as a consequence of its ability to explain reality, without conceiving of it as split into two irreconcilable domains, the new theory provides a logically superior account of sensory qualities in the world than the quality-based representational picture derived from Newton. What we need to move on to next, however, is the question of how the new theory might fulfil its promise and actually solve the problem of bridging the mind-body gap. That we will make our principle business in the next part of the book.

PART IV: CONSCIOUSNESS (SOLVING THE HARD PROBLEM)

18. The Visual Opening

The relational theory of sensory qualities has been shown to overcome all of the arguments that are traditionally held to disprove the idea that the physical world might be coloured and contain sounds and smells in an objective sense. We have also seen that because it offers a better account of the mind–body relation, it provides a logically superior account of reality than the quality-based representational picture.

So it seems we can now claim that the world is the way that our new theory describes it, rather than the way that it is described by representational theories. Thus when a flower blossoms unseen in the countryside it does manifest a quality of colour and when a tree falls to the ground in a remote forest it emits a crashing sound to its surroundings even though no-one may be present to hear it. However, this bold claim requires qualification, for we have not yet provided a full explanation of how observers become aware of the qualities of colour sound, smell and so forth which occur in the world.

MEAT OR MIND

The task of explaining how awareness of external sensory qualities is achieved raises a subsidiary question. For when a person uses his or her eyes to look upon the world, what is it that faces the glorious array of colours that are presented? More generally, for all sensory qualities presented from the world what exactly is the entity that accesses them in acts of perception? Is it just a complex lump of meat connected to an array of sense organs, or is there something beyond the merely biological going on inside an observer's head? Something even to which one might justifiably give the accolade of 'mind'?

This question represents a unique challenge for the direct realist outlook. While the quality-based representational approach is faced with the need to develop a plausible account of the mind, the direct realist conception of sensory qualities is confronted by an almost equally imposing hurdle of its own. It needs to develop an account of man that is capable of explaining how he perceives a world pregnant with sensory qualities perhaps *without a mind at all*. We know already that direct realism in general implies a virtually non-existent concept of mind, so one would expect our new theory to accept that if there is a mind it is of a minimal nature. In order to fully meet this new challenge, we must now flesh out the details of what that really means.

According to the new theory, surfaces present their colour qualities to the surrounding world whether or not observers are present. When a person comes along to view such a part of the world, they therefore gain 'access', as I have put it, to the pre-presented colours. But it is not possible to experience anything about the region inside our heads where such access occurs. Thus one sees no colours at the inner end of the visual field and one hears no sounds at the equivalent auditory position where hearing is conducted. In these places there are only transparency and silence. The Austrian-born philosopher Ludwig Wittgenstein tackled this point in his acclaimed work the *Tractatus Logico-Philosophicus*. Amongst the terse set of propositions which make up this tightly crafted account of things is an appeal to the fact that we do not see our own eyes within the visual field:

'5.6331 For the form of the visual field is surely not like this.'

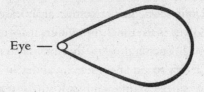

Wittgenstein (1922/1974)

Douglas Harding, a very different thinker, examined the same conundrum at greater length in the 1960's. Harding was an English writer who became famous for publishing a book called *On Having No Head* in which he recounted his Zen-like realisation, while resting on a ridge during a walk in the Himalayas that, as he puts it, 'he had no head' This is his account of the episode:

> 'What actually happened was something absurdly simple and unspectacular: just for the moment I stopped thinking... There existed only the Now, that present moment and what was clearly given in it. To look was enough. And what I found was khaki trouser legs terminating downwards in a pair of brown shoes... and a khaki shirtfront terminating upwards in – absolutely nothing whatever! Certainly not in a head. It took me no time at all to notice that this nothing, this hole where a head should have been, was no ordinary vacancy, no mere nothing. On the contrary, it was very much occupied. It was a vast emptiness vastly filled, a nothing that found room for everything – room for grass, trees, shadowy distant hills, and far above them snow – peaks like a row of angular clouds riding the blue sky. I had lost a head and gained a world.'[1]

Although neither Wittgenstein nor Harding were in any way attempting to defend direct realism, they both highlighted the same remarkable feature of man's visual experience. The inner end of the visual cone has no experiential content. It is at this inner end that our access to the colour qualities of the external world is gained yet it itself has no colour. It forms what might be called a 'visual opening' upon the external coloured realm. We also have access to other sensory qualities in the world but no awareness of smells, sounds and so forth 'in our heads' at the point of access to these qualities, so one can talk more generally about a 'perceptual opening' (the visual opening is most apparent because the geometry of the visual field is more obvious than that of our other sensory fields). So if there were to be a 'mind' rather than just a brain facing the world of external sensory qualities when we perceive it, then according to our new theory it would have to consist, at least in perceptual terms, of primarily this opening. Indeed, wherever the representational theory explains man's experience in terms of a mental representation composed of sensory qualities, the relational and direct realist account must do so in terms of this perceptual opening upon sensory qualities in the external realm.

Consequently, where a representational account of mind consists of the 'mega' concept of a quality-laden representation, our new theory becomes an account of the visual – or more generally the perceptual – opening.

But being devoid of all sensory qualities and, as Harding's account vividly illustrates, any other features that can in any way be characterised, such a 'mind-as-opening' is as good as non-existent. Furthermore, our new account expressly denies that a quality-laden representation including sensory qualities that mirror the external world is involved in acts of perception. So the new relational theory offers nothing else of which a perceiving mind could consist. In short, in stark contrast to the representational model's heavyweight conception, the mind which emerges from the new theory consists almost entirely of nothing. To all intents and purposes this outlook on perception suggests that there is no mind. It points to the conclusion that when a person is engaged in perceiving, what is facing the vibrant array of qualities which the world presents is not a mind-in-a-brain but just a brain alone.

19. Sensory Memory

So in a human act of perception the entity which faces the external world of brilliant sensory qualities is simply a brain, together with its sense organs, surmounting the observer's body. There is, as we argued in the previous chapter, no mind inside the observer's brain (or body). Furthermore, the objective nature of sensory qualities means that all of the sensory brilliance that a person becomes aware of through perception is presented to his location from external objects whether or not he is observing them. This raises the question of what contribution, if any, the complex neurology of the brain makes to the perceptual act. When all of the most fundamental ingredients of perceptual experience – the sensory qualities – are located in the external world, the only function required of neurological processes is that they somehow enable the access to or 'opening' upon this quality-rich world which perception offers. What conceivable process within the human brain could achieve this?

SENSORY MEMORY

The most likely candidate appears to be one known as 'sensory memory'. Sensory memory is the most fundamental of the forms of memory recognised by cognitive psychologists. In *Psychology: The Science of Behaviour* Carlson and Buskist (1997) describe it thus:

'Sensory memory is memory in which representations of the physical features of a stimulus are stored for a very brief time – perhaps for a second or less. This form of memory is difficult to distinguish from the act of perception. The information contained in sensory memory represents the original stimulus fairly accurately and contains all or most of the information that has just been perceived. For example, sensory memory contains a brief image of a sight we have just seen or a fleeting echo of a sound we have just heard.' [1]

According to the widely used 'multi-store' model, memory is made up of three 'stores', these being the sensory register, short-term memory and long-term memory.[2] The term 'sensory memory' refers to the sensory register, of which there are usually held to be one for each of the modes of sensing (thus a visual register, an acoustic register, an olfactory register and a haptic register and so on[3]). Information stays in sensory memory for approximately a fifth of a second[4] and then it, or at least that fraction of it which is attended to, passes on to short-term memory[5] from where it can flow to the long-term store.

We can best see the significance of sensory memory if we imagine a person who does not possess it (perhaps as a result of having had it temporarily disabled by a surgical procedure). We will put him in a room which contains a number of brightly coloured objects; say a collection of children's toys, strewn over the floor. Our theory states that the toys present their colours across the space of the room in advance of his entry, but what happens when such a person positions himself in the middle of the room and starts to look around? As his gaze shifts from brightly coloured toy to brightly coloured toy, each colour presented to the location which he occupies is simply replaced by another, which in turn is overtaken by the next and that by another, and so on. Everything is lost in the flux of the present moment. There is no temporal depth to the occurrence of the succeeding colours at his location. It is almost as if the person doesn't exist there. But now let us ask the surgeons to give him back his sensory memory. Once again the person returns to the toy-filled room. This time things happen quite differently. The colours endure for him in his immediate awareness. Thus, after his attention has moved from say, a red toy to a blue one there still remains a sense of the redness for a fifth of a second in his sensory memory. This time the first colour is not obliterated entirely by the second, and likewise as his gaze moves on to other colours. What sensory memory achieves, then, is to establish a basic continuity in the series of sensory quality presentations which occur during perception. I suggest that the term 'experience', in the context of perception and sensation, refers to nothing other than this ongoing sense of continuity.

Furthermore, because sensory memory is a process which goes on inside the brain of the perceiver, the sense of continuity which it produces is inevitably centred on that organ. I therefore also suggest that it is really this 'centre of continuity' that one is referring to in the notion of a visual or perceptual 'opening' upon reality. As we have seen the 'opening' does not itself have any sensory quality characteristics, and this is consistent with the idea that it is the brain's own process of sensory memory at work.

MAN, THE MEMORY MACHINE

So, sensory memory explains two things. It provides an account of what we call 'experience'. It also explains man's sense of an 'opening' upon reality as being the centre of his sense of sensory continuity. But on top of that psychologists argue that one function of sensory memory is to 'hold information long enough for it to be transferred to the next form of memory, short-term memory.'[6] As we have seen, according to the widely used multi-store theory, short-term memory in turn feeds into long-term. This means that an explanation of experience and the perceptual opening in terms of sensory memory integrates well with contemporary thinking about man's overall memory structure. Indeed, on this basis it could be said that when in an act of vision he faces the vibrant array of colours which a portion of the world presents to him, man amounts to nothing more than a biological memory machine. As we have seen he is not a mind. He is simply a physical body and in that body is a brain whose sole function – as far as sensing goes – is to store in memory the patterns of stimulation which determine presented sensory qualities. Firstly, in the almost immediate register of sensory memory which gives man the sense of a 'stream' of experience and the centre of continuity which we call an 'opening' upon reality. Then, in the short and long-term forms of memory which permit storage for subsequent recall.[7]

This is clearly a considerably simpler vision of man's nature than most. Yet despite that it permits full comprehension of all of the major subjective ingredients of man's perceptual experience. All perceptual content is located externally to man in the form of sensory qualities in the physical world. The sense of a direct opening upon this content-rich world and of experiential continuity which man enjoys are both provided by sensory memory.

20. On Having No Mind

According to the account that has now been assembled there isn't anything inside our heads which warrants the designation of 'mind'. Instead, everything that arises subjectively during acts of perception is accounted for in other ways. The sensory qualities that we become aware of are located externally in the physical world. Such other features as might be taken to be aspects of 'mind' – like the continuity of man's sensory experience and the visual opening (more generally the perceptual opening) at the inner end of the sensory field – are accounted for not by the strange entity we call 'mind' but by the physical and neurological mechanism of sensory memory.

What, then, are these supposed things called 'minds'? It is not possible to touch them, see them or interact with them externally in any way whatsoever. Furthermore no-one has been able to specify their nature in anything other than negative terms. (That is, they do *not* have mass, they do *not* have length, they do *not* have colour, and so on.) I suggest therefore that 'minds' are in fact a figment of man's collective imagination.[1] There simply are no such things.

ROOTS OF MIND

The idea that man has a 'mind' has deep intellectual roots, going back at least as far as the Ancient Greeks. But this long-standing concept of mind is of an entity which has wide-ranging capabilities encompassing many non-sensory capacities, including thought, imagination and emotion. The conviction that the minds we possess have as their primary function the task of constructing sensory qualities is more recent. It arises from the intellectual monopoly that has been enjoyed for more than two centuries by the prevailing quality-based representational theory of perception. This theory takes the now well-established causal picture of the sensory process and, following on from Newton, puts forward the claim that sensory qualities are mental constructs within the brain. For there to be mental constructs, there must be a mind capable of constructing them. But the arguments set out in this book demonstrate that there is no justification for such a view. The claim that sensory qualities are mental constructs has no basis, either on the grounds of the causal picture alone or in the light of the various perceptual phenomena, such as illusions and refraction effects, which are conventionally held to disprove direct realism (and which we have examined at length in earlier chapters). That this is so is shown by the fact that the new relational theory of sensory qualities provides an alternative account of perception which is compatible with the causal picture and can explain all such perceptual phenomena. As it holds that sensory qualities are not mental constructs, there is no need to posit the existence of such a thing as a mind that constructs them. It becomes evident that the whole concept of a mind that constructs sensory qualities is entirely an *artefact* of quality-based representational theories of perception[2] (and could be jettisoned were one to abandon this type of theory).

MIND AS MEMORY

However, it could still be argued that man's undoubted non-sensory capacities indicate the necessity of maintaining some notion of mind. But even here I suggest that this need not be so. In principle, it is possible to understand all 'mental' capacities such as thought, imagination and emotion within the framework of a conception of the human organism as an entirely physical being endowed solely with memory.

For example, the capacity to imagine is clearly derived from, and in some respects restricted to, a store of memories. This can be seen, if nothing else, from the fact that it is not possible to imagine a colour that one has not previously seen, or an odour that one has not previously smelt.[3] This suggests that imagination is simply the creative manipulation of sensory memories, and that in this sense it is more an augmentation of the more fundamental process of memory rather than a wholly separate facility in its own right. Thinking and emoting are less tangible as mental capacities but even here it can be suggested that they operate in a cumulative fashion[4] by – as with imagination – building on memory-based stores of past attitudes and responses with an element of creativity. Certainly, it is hard to conceive of either thought or emotion occurring in the complete absence of an ability to register within memory attitudes and responses to events (although these functions may not require the ability to consciously recall such registered episodes). In a similar manner, even the archetypal 'mental' function of free will could not operate without a context of memory, which enables the agent to imagine an action prior to enacting it, based on memories of having performed the action on previous occasions.

These ideas have only been briefly stated here. However, they demonstrate how one might consider many of the capacities traditionally thought to compose 'mind' as augmentations of the fundamental process of memory (rather than as wholly original functions). In this way, they can be brought within a model of man as a biological memory machine, composed only of a body/brain with neurologically-instantiated memory processes.

The only remaining challenge presented by traditional notions of mind is that of bodily sensations such as pain and tickles, and especially the famous 'phantom-limb sensation' (the feeling of the ongoing presence of a surgically-removed limb that is often felt by amputees). Most bodily sensations yield to an explanation parallel to that for external sensory qualities. So the qualitative content of such sensations – the felt pain or 'tickle' – can be conceived of as a relational property of space in the same way that external sensory qualities are. Only here, the quality happens to arise (under normal circumstances) solely within the space occupied by the subject's body, and it is (non-causally) determined by the information which passes along the relevant nerves in his or her central nervous system, in the same way that external sensory qualities are determined by the information in external signals. The case of the phantom-limb sensation can then be treated as an example of brain-based (or possibly body-based[5]) memory. My hypothesis here would be that the original information pattern in nerve signals – or perhaps some other biophysical phenomenon of the body – which at the pre-amputation stage determined the 'normal' feeling of the limb to be presented is made to recur as a result of a memory-like process. The result is that the same quality is made to be presented from space (previously occupied by the limb) to the subject, just as occurs in the case of after-images and other 'illusions'.

SENSING MIND

It might be an option in principle for anyone not persuaded by such ideas to divide their conception of 'mind' into two. On this strategy one would hold there to be a sensing element to mind, that element responsible for the construction of sensory qualities – and a non-sensing element – that element responsible for all of its other capacities such as thought, imagination and emotion. Such a division would make it possible to accept that the 'sensing mind' does not exist while still claiming that the non-sensing element of mind does. However, while it may be possible to follow such a strategy in theory it would be difficult to carry it through in practice, given that one of the essential characteristics of mind is its unity. I therefore set aside this option and conclude that, considered as a single entity, the mind does not exist. What this means in practice, as we have seen, is that sensory qualities with all of their richness of content are located externally, and the means by which we access them – identifiable with man's continuity of experience and his perceptual opening – are entirely physical processes of sensory memory within his brain. There is therefore no 'mind' within the matter of the brain.

As far back as Chapter 4 we identified the micro concept of mind as being a primary benefit of the direct realist outlook, because of its potential to solve the mind body problem. Here, in the conclusion that there is no mind, we now see the culmination of that concept.

21. On Having No Mind-body Problem

So what are the implications for the mind-body problem of man's having no mind? Quite simply, if man does not have a mind then he does not have a mind-body problem. According to the picture that has been developed here, man consists solely of a body, including a brain, its sense organs and processes of memory. However there is no 'mind' within this structure and so there cannot be a problem over any relation between it and the body.

The idea that man is exclusively physical is not in itself new. 'Physicalist' depictions of man go back at least as far as Hobbes, a contemporary opponent of Descartes, and they encompass much of the behaviourist tradition in psychology which was influential during the first half of the 20th century. But these depictions have never simultaneously been able to reduce the mental comprehensively to the physical and provide an adequate explanation for that element of experience which is constituted by the sensory qualities. It is because of this failure that there exists the difficulty of accounting for how experience arises in the grey matter of the brain (David Chalmers' so-called 'hard problem of consciousness').[1] The core of the mind, or 'consciousness', is provided by sensory experience and, as Chalmers has pointed out,[2] there is a profound explanatory gap between such experience and any account of it that is presented in terms of the material functioning of the brain.

During the last hundred years or so there have been a number of ingenious attempts to develop entirely physical accounts of the mind. These have ranged from the simple, such as the 'mind–brain identity theory', which asserts that mental processes simply are brain processes, to the more sophisticated such as 'functionalism', based on an analogy with computers, according to which all mental states have inputs and outputs and are related to the functions of the brain in the same way that software is to a computer. But if there is one problem which all such ideas share it has been their inability to account for the sensory qualities.

MONISM WITHOUT MIND

The relational theory of sensory qualities that we have developed here offers a major contrast in this regard. For it not only depicts man as entirely physical but does so while providing a full account of the sensory qualities (as external physical relationships). It therefore manages to explain all subjective aspects of sensory experience entirely within the context of a physical depiction of man. In doing so, it provides an account of experience entirely in terms of the physical realm and thus solves the hard problem of consciousness. The consequence is that the old dualist metaphysics derived from Descartes and Newton is no longer required. For if conscious beings such as man are not divided into the twin realms of mind and body, then reality as a whole can be conceived of as a single domain.

22. The Hard Problem of Reality

Although our new theory rids us of the mind-body problem it could be argued that it does so only at the cost of shifting the problem out into the external world. Some might object that in the final analysis it has not managed to provide a detailed causal or 'mechanical' account of how changes in physical signals like light, sound waves and thermal radiation alter the corresponding sensory qualities that are presented across spans of physical space. The best that we have been able to do is to suggest that the information content in such signals affects the relations obtaining between 'source' and 'destination' locations. In turn, these relations determine the physical relationships which make up the corresponding sensory qualities (of colour, sound, heat and so forth) that are presented between locations. But these ideas hardly amount to a precise causal account of the dynamic interaction of well-defined microscopic particles or forces, of the type that is expected of contemporary scientific explanations of physical phenomena.

HARD AND GENERAL

The principal reason for this shortcoming is that our new theory suffers from a parallel problem to that which confronts the quality-based representational account. The latter claims that sensory qualities arise within the 'mind' and finds itself unable to explain how they are causally affected by the matter of the brain. Whereas, conversely, the new relational theory of sensory qualities (with its claim that sensory qualities occur in the external world) is unable to provide a causally-rich explanation of how they are affected by the physical matter of that world.

What this illustrates is that there is not only a 'hard problem of consciousness', concerned with how the sensory qualities we experience are brought into awareness by our brains, but also a generic 'hard problem of reality', having to do with how matter in general brings about changes in sensory qualities in the external world. Or, rather, that the hard problem of consciousness is a facet of this broader hard problem of reality. In the end, there cannot be any solution to the problem of how the causal realm – including both the inner matter of the brain and the outer matter of the external world – impacts on the sensory qualities.

The reason for this lies in the ultimately simple nature of the sensory qualities. This characteristic is well illustrated, as we have previously observed, in the fact that fields of sensory qualities, such as the colour of a rose's petals, and the aroma of freshly-ground coffee, have no internal elements. Sensory qualities are thus quite possibly the simplest entities in the universe. In their simplicity – and also homogeneity, as Wilfrid Sellars pointed out in what has come to be known as the 'grain problem'[1] – they are distinct from the composite or 'grainy' nature of the matter of atoms and molecules which constitutes the causal realm of the physical universe. There can therefore never be a detailed 'mechanical' account of how the physical realm (whether internal or external) causally brings about changes in such hyper-simple entities. There are no internal structural elements within sensory qualities with which the causality of atomic matter might interact. Hence there can be no effects of atomic matter on sensory qualities.

RESEMBLANCE AND REALITY
The existence of the hard problem of reality means that both the Newton-inspired representational account and our new theory of

sensory qualities suffer from an identical and impassable, explanatory hurdle.

Yet both schemes attain the same level of internal *coherence* and integration with scientific thought. In what respects, then, can either one be said to be superior to the other? In these circumstances it seems that the only basis left upon which to make a comparison between the two is the degree to which their accounts of things *correspond* to the way that the world in fact appears. When there exists two comprehensive accounts of reality, neither with an overwhelming case over the other, both equally coherent and both distinct from each other, then the ultimate arbiter between them can only be their correspondence to the appearance of reality itself. It would be wilful in the extreme, all other things being equal, to prefer a theory which described reality in a way that was in contradiction to the way that it appeared over one that was consistent with its appearance.

At this point the very 'naivety' of our relational theory becomes its greatest strength. For the appearance presented by reality is one in which sensory qualities seem to be inherent features of the external environment. While our new conception of sensory qualities is consistent with this aspect of reality's appearance, the quality-based representational account directly contradicts it by depicting sensory qualities as mental constructs. The latter therefore couldn't be more at odds with this aspect of the appearance presented by reality. It would be unreasonable to prefer as the best account of reality the quality-based representational theory which depicts a version of reality completely different from the way that the world in fact appears. Thus the ultimate grounds on which our new theory has logical superiority over its rival are quite simply that it corresponds to the way that reality appears. The conclusion must be that of the two it is the new relational theory of sensory qualities which provides the true description of reality.

PART V:
CONSEQUENCES

23. A World of Value

If the conclusions presented so far are correct, it is almost inevitable that the various consequences which follow from them will seem startling. Except for the 'selective' theory of Edwin B. Holt which is in the final analysis incomplete,[1] the notion that sensory qualities exist externally has rarely been given serious consideration as a *physical* thesis during the entire period of modern philosophy. This means that as a concept it is alien to contemporary modes of thought. Its ramifications are therefore unfamiliar to the majority of people who accept the world view derived from the prevailing account of perception, together with its assertion that sensory qualities are mental constructs.

A good illustration of just how unaccustomed modern man is to those consequences is provided by a 'mystery' that appears to be raised by the relational theory. On the view that sensory qualities exist objectively, in the absence of observers coloured surfaces such as those of cars, buildings and mountain-sides must present their colours to the external world, but no subject need be 'out there' in the world experiencing them. This seems a peculiar idea, because it is hard for those of us accustomed to contemporary ways of thinking about perception to accept that appearances can present outwardly and yet there be no conscious observer experiencing them at the point to which the presentation is made.[2] However, there is nothing about sensory qualities which makes it intrinsically impossible for them to manifest to observer-free locations, so there is no real mystery here. The fact that we find it peculiar only illustrates the hold which prevailing modes of thought have over us and the power of the association between sensory qualities and subjectivity that they generate. The phenomenon of presentation to observer-free locations does, however, demonstrate the extent of the difference in our understanding of sensory qualities that is brought about by the relational theory.

METAPHYSICS OF APPEARANCE

The presentation of sensory qualities to subject-free environments is only one of a large number of novel issues which the relational theory of sensory qualities generates – many of which we have encountered in previous chapters. What they collectively point to is the long-term need to develop a metaphysical account of reality based in direct realism. This would be an integrated conceptual world view similar in scope to that of the metaphysics that we currently live under (derived from the representational theory of perception), in which such concepts as presentation to subject-free environments, sensory qualities being properties of space, and there being no minds would each find their place.

Because sensory perception acts as a bridge between man and the external world such a new version of metaphysics would undoubtedly have things to say about both the nature of man and of the external world.

However, I will not attempt here to complete the assembly of such an entire metaphysics. Instead my aim will be limited to examining the ramifications of the new relational account. In the current chapter the focus will be on the implications of that account for our understanding of the physical realm, while in the final chapter I will turn to consider its impact on our understanding of man. However, during the investigation I will indicate, as they are encountered, any points that seem to be of significance in relation to the development of a direct realist metaphysics in the future.

MEANING OF ART

The consequences of the relational theory for our concept of the physical realm derive from its claim that sensory qualities exist objectively in the external world. (This contrasts with the quality-based representational outlook, which can be taken as implying that the physical realm is devoid of sensory qualities.)

For human beings there could hardly be a more significant question regarding the physical universe. This is because a world in which sensory qualities intrinsically exist – one such as that envisaged by our new account – is one which bears meaning to man. Whereas a world devoid of sensory qualities – as envisaged in the view associated with the quality-based representational theory – is one that is meaningless to man. This can be looked at in a variety of ways, but we will do so first of all by thinking of art.

Each of us knows at least one painting and one piece of music which is redolent with meaning. For many, outstanding pieces of music, for example by the 'great' rock bands such as Pink Floyd and the Rolling Stones, offer incontrovertible examples of the latter. Yet when a rock band like the Stones or Pink Floyd play live, the aural results can be construed in two distinct ways in accordance with the interpretation of sensory qualities put forward by the rival theories of perception that we have been considering.

Firstly, it is possible to take the view associated with the quality-based representational theory that, objectively-speaking, the efforts of such a band produce no sensory quality of sound in the physical environment of the auditorium in which they are performing, because the speakers to which their instruments are attached generate only sound waves in the physical realm. On this view, in terms of sensory qualities, there is no more sound quality produced (although many more sound waves) in the space of the auditorium than if the band members had been playing air-guitars. Or alternatively, it can be considered, as proposed by our new theory, that intonations of the sensory quality of sound *are* brought about in the auditorium along with the sound waves. What impact do these opposing interpretations have on our understanding of the meaning that is felt by the audience to reside in the band's music?

In the first case, sound waves are only ripples of increased compression between air molecules that travel through the atmosphere, so it is difficult to see how the meaning that so strongly appears to be 'in' the music could in any way be attached to them. If the only thing of a musical nature that truly exists in a concert hall when the Rolling Stones play are such disturbances of pressure in the air, then the meaning discerned in the artists' music can hardly be intrinsic to the sound waves alone. One would have to suppose that the meaning apparent within the music is in fact manufactured in some way by the listeners' mental faculties. But matters are quite different on the realist view that the *sensory quality* of sound genuinely exists in the physical realm. Here it becomes entirely conceivable[3] that the meaning detected in the music could be intrinsic to the events occurring in the space of the auditorium and not a matter of merely being imposed by man's mental faculties.

This fundamental difference gives an indication of the profound link between sensory qualities and meaning, a point which is reinforced when we consider another form of art, painting. The Impressionist works of Claude Monet, such as his celebrated paintings of his water lily garden, are famously dependent on the quality of colour for their meaning. Even photographed in black-and-white, a water lily garden painting by Monet loses much of its original meaning to the viewer. So, what fate does a Monet painting face, when considered on the interpretation which claims that the sensory quality of colour does not genuinely exist in the external world? On this view, the painting reduces in an objective sense to little more than a rectangle of canvas patterned with daubs of dried, pigmented oil which possess no sensory quality of colour. In this colourless state how much meaning can the painting possibly express? If a black-and-white Monet expresses only a diluted version of the original's meaning, such a piece of fabric, lacking all colour quality, would be devoid of all meaning.

The point of principle brought out by these examples extends beyond the work of famous artists and carries through to all music and painting. Sensory qualities are fundamental to the ability of works of art to express meaning. Without the sensory quality of sound, a work of music is, considered as an objective entity, a meaningless package of sound waves. Likewise, a painting conceived as drained of its sensory quality of colour becomes a mere rag coated in chemicals. Art in its various forms supplies humans with meaning in a raw and direct fashion and yet its possession of sensory qualities is an essential prerequisite to its being able to carry out that function.

ART OF THE UNIVERSE

Another way to see the dependence of meaning on sensory qualities is to imagine a universe without them. Or equivalently, to imagine the actual universe as it is held to appear according to prevailing ideas about perception, which are associated with the idea that sensory qualities are solely mental constructs.

It is possible that no coherent account of the objective appearance of reality lies embedded in the prevailing account.[4] However, in so far as an intelligible view can be extracted, it seems to be that the intrinsic appearance of the physical realm is that it is silent, colourless and lacking in all qualities, such as smell, warmth and taste. As transparency forms the primary example known to man of being 'colourless' in the case of macroscopic physical objects (as opposed to microscopic ones or abstract entities), this suggests that the objective appearance of the everyday physical realm is to be taken as being one of boundless transparency and silence, together with the absence of all other qualities. 'Boundless' transparency because in the case of objective reality without qualities, there could be no instances of colour to limit the 'clearness' as occurs in normal vision where the transparency of the atmosphere reaches only as far as the coloured surfaces at its limits. Thus all material structures would have to be transparent, and clearness would extend to the furthest reaches of the cosmos. As I say, it is questionable whether such a position makes sense (something which should perhaps be of concern to anyone who sought to defend the quality-based representational position) but the point for us here is simply that such a universe devoid of all sensory qualities, has also *lost all meaning*. This can be seen if one imagines walking through a stretch of countryside twice. Once, in the normal version of the universe, and taking in the full panoply of colours, sounds and smells presented from the nearby fields as one strolls along while feeling the warmth of the sun beating down on one's face. But then again while only able to perceive the scene as it would appear according to the quality-depleted view associated with current thinking about perception. On this second occasion the trees, grass, hillsides and the sky as far as one could see would all appear transparent. There would be no colours anywhere. The bird-song and burbling of a nearby stream that were audible on the first occasion would both be lost, as would the smell of sheep and of cow-parsley in the fields. There would be utter silence and a lack of odour, in a limitless expanse of transparency. All of the meaning expressed by the world on one's previous encounter with it would simply have evaporated in raw uniformity.

What one sees here is that just as with paintings stripped of their colour, and music lacking in sound-quality, our universe if stripped of its sensory qualities would also be robbed of all of its meaning. They are the foundation of all variety and content and thus the source of all meaning.

NEWTON'S SLEEP

It may have been an appreciation of the fundamental importance of the sensory qualities in this regard which led the poet William Blake to his sustained attack on Newton's philosophy. He said, 'May God us keep from Single vision & Newton's sleep.'[5] Here Blake may have been referring to the human feelings of emptiness (or 'sleep') that are prompted by the uniformity of the material world as understood under the Newton-inspired account of perception. But possibly his most potent attack took a graphical form, in an engraving entitled simply 'Newton'. In this he showed the scientist as a god-like form in embryonic pose crouching on, and emergent from, the material bulk provided by a great boulder in the approximate form of an egg. The figure of the scientist uses a pair of dividers while studying with single-minded intensity an abstract problem of geometry set out before him on a roll of paper.

Blake's engraving has been the subject of diverse interpretations, but the one I commend is that his commentary on the work of Newton is symbolised in the engraving by the colouring which surrounds the central figure of the scientist in the image. As a visual artist Blake's principal concern would have been that Newton's perceptual theory implied, or has been widely taken as implying, that the material world was objectively without colour, and this is represented, I suggest, by the region around the figure of Newton, hard at work on his mathematical and scientific abstractions, being only uniformly and darkly coloured. It is as if the world as it is conceived in accord with the great thinker's scientific ideas is without human significance and has 'gone to sleep'. But in contrast, the other areas of the image where the extremity of the great boulder lies – representing, perhaps, by its solid bulk, and the 'birth' of the scientist from its mass, the bedrock of the actual substance of the physical universe – there is a wealth of colour. So on this interpretation what Blake's engraving conveys, with considerable insight on the poet's part, is that the physical universe actually possesses sensory qualities such as colour but that the scientific work of Newton leads to an understanding in which it is stripped of these qualities and a state of uniformity and loss of significance, analogous to the darkness of sleep, is arrived at.

ATOMS OF MEANING

Such is the importance of the role of the sensory qualities in conveying meaning that it is possible that a future direct realist metaphysics could see them as being foundational sources of meaning – as carrying out the equivalent function for meaning that atoms carry out for matter. Yet another situation where their fundamental significance in this regard is made apparent, is on those occasions when we realise that it is impossible to describe sensory qualities. This limitation is often expressed in the remark that two people cannot know whether they are experiencing the same colour, sound or odour quality because they cannot describe the quality to each other. The reason why sensory qualities cannot be described in this way has much to do with their simplicity and lack of internal elements. But it is also, perhaps, an aspect of their being raw features of meaning in the universe. Thus it is hardly to be expected that meaning itself should be open to description.

FOUNDATION OF VALUE

Whether the meaning that sensory qualities make available in the universe is an objective feature of the external realm, or solely a human response to external sensory qualities, I leave as an open question to be resolved in a future metaphysics. But whatever its precise standing, it is, I suggest, the foundation of all judgements that the physical realm has value. Thus it is not possible to discern any value in the uniform and barren void of the quality-depleted universe associated with the representational theory of perception. Whereas with the return of sensory qualities to the physical realm brought about by our new relational theory of sensory qualities such value comes flooding back. A world full of colour, sound, smell, taste and warmth has richness of content of a type that it is possible for sentient beings such as humans to respond to in a profound way.

COLOUR AND CAPITALISM

It is even possible, it seems to me, to argue that on the large scale, Western man's actions in the physical environment have been subject in recent history to two major influences. On the one hand, Adam Smith's capitalism[6] has provided the 'motor' which has driven behaviour. But on the other hand, the context within which this frenzy of activity has occurred has been provided by the dismissive set of attitudes to the physical environment (as having no intrinsic value) derived from the perceptual theory of Newton, Locke and Berkeley. The utilitarian response to the physical realm made possible by the associated view of the sensory qualities, has arguably played a significant role in enabling society's capitalist motor to bring about unprecedented levels of commercial activity because it has meant that there has been no requirement to respond to any aesthetic or other value-based concerns regarding the external environment.

The reintroduction of sensory qualities to man's concept of the physical realm that occurs under the new theory could potentially change all of that. The 'motor' of capitalism shows no signs of abating in its central role as provider of the driving force behind much human activity. But an acknowledgement of the existence of sensory qualities in the physical realm brings with it acceptance of the fact that it is a domain that is pregnant with meaning and value. I suggest that this might constrain the capitalist 'motor' and adjust social activity broadly through two mechanisms.

NEW PRIMITIVISM

In the first place, much as for primitive man (who has no Newtonian theory of perception and is therefore untroubled by the notion that sensory qualities might not be in the external realm), there would cease to be a disjunction between man and his aesthetic response to the meaning offered by the physical world. That response is one which is in general aesthetically positive. This can be seen by once again starting out from the case of art, where beauty, it can be argued, is simply the positive valuation of the meaning which art offers (while ugliness is its negative valuation). Clearly, from the examples that we have considered, comparing the rich universe of sensory qualities, which follows from the relational theory of sensory qualities, with the blank and uniform void associated with quality-based representational theories, man will give positive valuation to the meaning offered by the universe's sensory qualities compared with its absence. In that sense, then, it can be said that man's response to a universe replete with sensory qualities is an aesthetically positive one. (It could even be argued that such a universe is in general beautiful – a point of note for a direct realist metaphysics).

This positive aesthetic response to the sensory aspects of the world has always been present for man but a direct realist account of the sensory qualities changes the way that it would be interpreted. Thus the new theory does not affect the way that man actually sees but it does modify his understanding of the world seen (and of himself the seer). This change is such as to suggest that his positive aesthetic response to physical reality is an accurate and genuine one, because the apparent colours, sounds, smells and so forth on which it is based *are out there*. So this puts modern man more into the sort of unsullied positive aesthetic relationship with the physical realm that primitive peoples enjoy, in which the apparently positive nature of the relationship is accepted as having a sound basis.

I suggest that the practical impact on human behaviour (as driven by the motor of capitalism) of an acceptance of the integrity of this positive aesthetic response could be substantial. Taking man's endeavours in constructing his built environment as an example, it could conceivably translate into a response somewhat parallel to that of primitive man where every available surface becomes coloured and decorated. We might move towards a bright and colourful modern architecture quite alien to the uniform and brutalist structures that man has had to endure under the utilitarian influences of prevailing theories of perception.

Deeper Green

Another area of potential impact is in the field of attitudes to the environment. In current circumstances, where utilitarian motivations regarding the physical world are at the fore, concerns about the quality of the natural environment have a lowly place in human affairs. It is generally the case that unless they can be shown to have human benefits, policies that would make a positive contribution to the state of the environment struggle to gain acceptance in the political arena. (Indeed it could be argued that a contributory factor in the destruction of the earth's environment during the industrial age has been the negative – or at best neutral – attitudes to the physical realm brought about by the dominant theory of perception.)

With a return to a sense of value in the physical realm brought about by the reintroduction of sensory qualities this could change. For if at root man begins to value his physical context rather than dismisses it as meaningless then he is likely to give at least equal significance to its condition as to the purposes to which it can be moulded. Simply through a philosophical switch from a representational account of the sensory qualities to a direct realist one we might therefore see a change in underlying human attitudes to the value of the natural environment as an objective, physical realm. This might be sufficient potentially to transform peoples' decisions concerning the treatment of the environment. It is even possible to conceive that in a direct realist metaphysics of the future, the condition of the physical realm (as source of meaning, in the form of the sensory qualities) might be given the status of a moral good. If that sounds alien to our contemporary ears, it is only because we are so unused to thinking outside the boundaries of the all-encompassing metaphysics imposed by the representational model of perception.

The consequences of the relational theory of sensory qualities for our understanding of the external world have turned out, then, to be substantial. In sum, they elevate its significance by returning meaning to it and thus generating a human sense of value in the physical realm. However, as we shall see in the next chapter, this occurs at the cost of a significant downgrading to the status of man himself.

24. The Function of the Neuron

We now turn to the consequences of the relational theory of sensory qualities for our understanding of man.

END OF ALIENATION

A first and very direct impact derives from the new account's suggestion that the experiences made available to man through perception give him access to the world's colours, sounds and smells as they objectively exist, and are therefore in essence truthful. This is in marked contrast to the viewpoint of the quality-based representational theory, that sensory qualities do not exist objectively in the external world, and that the appearance of the world accessed through perception as coloured, sonorous, odorous and so on, is at its core a deception. The distinction has considerable significance because sensory perception creates for man an intimate relationship between himself and all that exists externally to him. The integrity of that relationship is dependent on the veracity with which what exists externally is understood to be represented. As one of the most fundamental, and perhaps *the* most fundamental, of all of the relationships in the life of man, this in turn also affects man's own sense of integration with reality.

Compared to the fluency with which animals inhabit their environment, post-industrial (and hence post-Newtonian) man has demonstrated a considerable discord in his relationship with the physical realm. Existentialism, which is concerned primarily with the uniqueness of each human individual's existence (knowledge of which comes notably through sensory experience) has put names to some of these 'existential' afflictions; including 'angst', 'alienation' and 'inauthenticity'. But what is their cause? I suggest that because of its negative impact on man's key relationship with reality, one can lay at least some of the blame for the ill-ease modern man feels within the physical realm – a phenomenon that is notably absent in primitive societies and amongst animals, neither of which consider sensory qualities to be anywhere but external – at the door of the quality-based representational theory of perception. The roots of the discord may be the representational theory's suggestion that man's central relationship with reality is deceptive. Correspondingly, by restoring the idea that perception delivers a truthful access to the content of reality (i.e.: its sensory qualities), the new relational theory of sensory qualities may undercut the basis of such existential discontents. It is therefore conceivable that under the influence of the new account, man may be able to return to living with a new ease in the physical domain, to a state similar to the pre-philosophical fluency with which animals and primitive peoples inhabit the world.

CELLS OF MEMORY

In terms of our understanding of man, most of the other consequences of the new account stem from its reinterpretation of what it is that occurs inside a perceiver's head during an act of perception. As we have seen, this involves the notion that there is no mind within the brain but instead only processes of memory.

The significance of this is that it implies the need to revise our understanding of the role which neurons play in sensory processes, and thus to redraft much of the explanatory basis underlying the neurosciences. At present, the conventional picture of perception implies that the role of nerve-cells in the sensory pathways of the brain is to contribute to the building up of a quality-laden representation of the external scene. There are known to be groups of neurons in the brain which respond to elements of a visual scene such as colour, edges and even higher-order features like facial movements.[1] Some neuroscientists hold that somehow these responses build up to a quality-laden representation of the external world. For example, this form of approach is exemplified in the recent work of Gerald Edelman,[2] in which he argues that consciousness is constituted by the integrated working of a core but variable set of neuronal groups within the brain. Sensory qualities (or 'qualia' as Edelman terms them) such as the colour red are the activity of this entire 'dynamic core', when it includes processing by a group of neurons that responds to light of 'red' wavelength.[3]

The relational theory of sensory qualities, however, undermines the need for this form of explanation of the role of neurons in the sensory pathways of the brain. The neurons no longer need to be imagined to contribute to the 'generation' (as Edelman puts it) of qualia within the interior of the brain, in such a manner as to form a quality-laden representation of the external scene. Instead, the function of neurons becomes that simply of registering in sensory memory – and subsequently longer-term memory – the information contained in sensory stimulation. This is not a case, note, of memorising the *appearance* of an external scene; the appearance occurs as an independent matter in the form of relational sensory qualities that occur externally. Neurons only have the task of registering in memory the information content associated with those qualities and encoded in signals such as light and sound-waves and subsequently in patterns of neural excitation.

On the new account, all of the sensory qualities, with their richness and variety of content, occur in the external physical domain and none within the confines of the brain. Therefore, there is no requirement for neurons to have any dealings with sensory qualities at all, and certainly not to 'generate' them. Instead, the cells of the brain have the much simpler task of operating with the electrical signals that pass to them from the perceiver's sense organs following stimulation by external, physical signals (or else generated within the brain itself, in the case of illusions such as hallucinations). They only have to register the information contained in these electrical signals in one of the levels of memory store (sensory memory, short-term, medium-term, etc.) for what we call 'experience' to occur.[4]

The new interpretation applies to the functioning of neurons in all acts of perception. This includes normal, 'veridical' perception and also illusory forms of perception – not least, those specialised illusions that are investigated by scientists in order to study the operation of the perceptual system. As an example of its application in the former case, when a person sees a red tomato the colour quality of the fruit occurs entirely on its surface (as redness-presented-to), and is presented across space to the surroundings. 'Red' light of 650 nanometres that has been reflected from the tomato's skin and embodies information about the skin enters the viewer's eyes. This gives rise to a pattern of electrical signals in the optic nerve and subsequently the visual cortex which also encodes the same information. The task of the neural cells is *simply to register this information in memory.* It is not to recreate the red sensory quality in the form of a little image deep inside the brain. (There being no need for this as it occurs externally on the tomato.)

As examples of the case of illusion consider the following images:

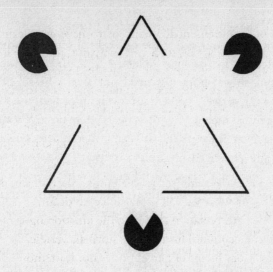

An illusory triangle

The 'vase–face' illusion

In general, the information registered in memory by neurons about both the colour and shape of external surfaces is the same as that encoded in the light reflected from them. However, as the illusions above show, this need not always be so. Neurons themselves can actively contribute to the information contained in the perceptual stream. Thus the visual system positively attempts to perceive form in scenes and this can lead to it detecting illusory contours and enhanced brightness as occurs when you look at the image involving triangles and circles. What almost certainly happens in this case is that those neurons which specialise in registering in memory any edges detected in a visual scene are 'duped' into firing by a combination of the visual system's propensity for seeking out shapes in scenes and the shape of a triangle suggested in this image by the incisions in the three bold circles. Hence information concerning the edges of a triangle that does not in fact exist in external reality is registered in memory and the viewer experiences its apparent triangular 'surface' as slightly brighter than the background.

It is often borderline in such illusions whether a distinct sensory quality appears as presented from the illusory image. There are even cases, as with the familiar 'vase-face' image on the previous page, where the 'brighter' parts of the pattern appear to oscillate (in this case between two faces coloured black against a white background and a white goblet against a black background). In the new interpretation there are two ways of understanding such effects. Where an illusory sensory quality is apparent then the relational theory of sensory qualities proposes (as we saw in some detail for after-images and hallucinations in Chapter 16) that it is a property of space determined as a result of the information content of the viewer's visual system. In those cases where no sensory quality is presented then the illusion can be interpreted as resulting from heightened registration of edge and brightness information in memory, but not at a sufficient intensity to bring about the determination of an external sensory quality.

The direct realist interpretation of the function of the neuron takes in cases both of veridical and of illusory perception and encompasses the entire field of perception. All known neurological responses to sensory stimulation can therefore be reinterpreted in its terms. Yet while it changes everything in the neurosciences, at the same time in a sense it changes nothing. The empirical facts of how neural cells respond to incoming sensory stimuli are left unaltered (being accepted as they stand) and no new predictions are made. It is only the understanding of what goes on when a neuron in an organism's sensory system fires in response to stimulation that is modified. Instead of contributing to a process of generating 'mental' sensory qualities (a process that is in principle incapable of explanation) they are understood as merely registering patterns of information in memory (and explaining this process only warrants the rating of an 'easy' problem of consciousness).

THE COMPLETE BRAIN

Reinterpreting neuronal functions in terms of the registration of information within memory represents a considerable simplification of the philosophical basis of the neurosciences. It opens up the possibility in principle of achieving a *complete* account of the brain. This follows because memory is a tangible, comprehensible phenomenon that is subject to investigation by experimental procedure, whereas the 'generation of sensory qualities' is a mythical notion that no scientific instrument will ever be able to probe. It follows that the neurosciences might eventually be considered an area of science that could in principle become conceptually complete, just as, for example, we conceive that chemistry could be.

THE LOLLY-POP STICK COMPUTER

There is a further advantage also offered by this substantial reinterpretation of the role of neurons in the sensory pathways of the brain. For it takes the force out of a paradox which arises from current ways of thinking about the mind. It is a common view amongst philosophers at present[5] that there is nothing unique about the physical realisation of man's mental faculties in the biological processes of his brain. According to this idea, if processes with identical functionality arose, or were constructed, in a different physical medium, then the same mental functions would arise. This is held to be true even if the medium was radically different. For example, one such as a silicon-based computer, or, in principle, an enormously complex mechanical assemblage made of wooden components like lolly-pop sticks, and rubber-bands. While this is almost impossible to accept when one's concept of man's mental faculties includes the notion of a quality-laden representation (how could the colour quality green, for example, arise amongst lolly-pop sticks and rubber bands?), it is much less challenging when one is only demanding that memory be instantiated in the device. We already know, for example, that memory occurs in silicon-based computers and it is not entirely inconceivable that a device made from billions of interacting wooden and rubber-band components could also be provided with some sort of memory function.

REPRESENTATION AND REALISM

Despite its 'direct' nature the relational theory of sensory qualities should not be taken as implying that the concept of a representation has no part to play in the new understanding of the role of the neuron. On the contrary, the existence of regions of the cortex which map areas of sensory organs such as the retinas of the eyes is well-attested to by experiment. But how such areas are conceived of necessity changes under the new account. Instead of being possible stages in the assembly of a representation which contains sensory qualities, they become simply a spatially-organised form of memory store (or stages in the assembly of such a store). So we see here that the concept of a perceptual representation can continue to survive within our interpretation of neurological functionality, but as a structure of memory not of sensory qualities.

A SLIVER OF REALITY

Finally, let us take a step back from both man and the external world to look at the consequences of the new relational theory of sensory qualities. If we do so we observe that the impact of the new outlook in the two areas balances out. On the side of the external world the physical domain has its significance greatly elevated as a result of its meaning and human value being restored to pre-philosophical levels. But on the side of man there is a considerable downgrading of status, as all human pretensions to mentality and distinction from physical reality are undermined. In many respects this adjustment could be seen as part of a long-term historical process in which the human sense of being 'unique' has gradually been eroded through the results of man's own attempts to understand his context – as exemplified by the abandonment in the Middle Ages of the idea that the earth lay at the centre of the cosmos, and later, Darwin's discovery of man's evolutionary descent from the apes.

***

It is possible now to state that the 'something' which experience implies must exist is physical reality. In other words it is a single domain reality which includes what we normally think of as the physical realm but not the mental. Contrary to prevailing thought, the sensory qualities which make up experience are located within it as relational properties of space.

***

NOTES

Introduction

Part I: Colour Unconfined

1. BLUE SKY THOUGHTS

1. *Light & Colour in the Open Air,* Minnaert, pg. 239.

2. *Mysticism and Logic,* Bertrand Russell, pg. 140.

3. *The Feeling of What Happens,* Antonio Damasio, pg. 9.

2. FOREST OF THE SENSES

1. Hirst is notorious in the art world as something of an *enfant terrible,* a reputation acquired from early works such as his 'The Physical Impossibility of Death in the Mind of Someone Living' which famously consisted of a dead tiger shark suspended in a glass tank of formaldehyde.

2. In the standard model of subatomic particles, quarks are a fundamental particle. They make up the neutrons and protons which form the nuclei of atoms.

3. DESCARTES' PERCEPTION MACHINE

1. *Meditations,* René Descartes.

2. *Descartes,* A.C. Grayling, pg. 125.

3. According to Beeckman wet bodies were those composed of round atoms whereas dry bodies were those composed of 'sharp' atoms (see 'Atomism in the 17th Century' from the *Dictionary of the History of Ideas, Studies of Selected Pivotal Ideas* edited by Philip P. Wiener).

4. Descartes went through a highly adventurous phase as a young man when, preferring not to follow the family tradition of a career in law, he enlisted in the army during the early phases of the Thirty Years War. He also travelled widely throughout war-torn Europe sometimes crossing militarily uncertain territory. It was during this period that he met and corresponded with Beeckman, who was one of the pioneers of mechanistic physical theory based on the new atomism. While staying in Paris, Descartes undertook experiments in optics in which he became one of the earliest to demonstrate the law of refraction of light (and subsequently the first to publish its mathematical description). (See *Descartes* by A.C. Grayling.) But it was in Amsterdam, where he eventually settled, that he made his most significant contributions towards the growing consensus in favour of a mechanistic and atomistic account of the physical world.

Anatomy was a field that Descartes may not have been wholly unfamiliar with as he came from a family with a tradition in medicine (and law), and in Amsterdam, whether by design or good fortune his lodgings were on a street of butchers – the perfect spot for collecting animals' body parts for dissection.

Descartes reports in correspondence visiting the butchers on a daily basis in order to observe the slaughter of cattle and to take parts that he intended to dissect back to his lodgings. It is clear from the detailed anatomical knowledge which he subsequently displayed in writings that came out of this period that he fully immersed himself in such efforts. Certainly his lodgings cannot have been a pretty sight at times. One imagines a table covered with stained cloth, buckets brimming with blood and trays full of dirtied implements. This picture is corroborated by a story told of Descartes being visited by a friend and asked to see his library. The man was taken out to a shed and shown the carcass of a dead calf hanging ready for dissection (the point of the anecdote being to indicate that Descartes' studies were, as he himself was to put it, directed at the 'book of the world').

5. *Cogito Ergo Sum, The Life of René Descartes*, R. Watson, pg. 15.

6. *The World and Other Writings*, René Descartes, pg. 146.

7. As above, pg. 146.

8. As above, pg. 117.

9. As above, pg. 148.

10. As above, pg. 149.

11. As above, pg. 3.

4. NEWTON AGAINST THE NATURAL VIEW

1. *Never at Rest: A Biography of Isaac Newton,* Westfall pg. 157.

2. As above, pg. 90.

3. As above, pg. 94.

4. As above, pg. 156.

5. As above, pg. 143.

6. *Theories of Light From Descartes to Newton,* Sabra, pg. 241.

7. *Opticks,* Newton, pg. 123.

8. As above, pg. 124.

9. As above, pg. 124.

10. As above, pg. 179.

11. As above, pg. 179.

12. As above, pg. 124.

13. As above, pg. 346.

14. As above, pg. 345.

15. The *Opticks* could be considered one the earliest works of 'popular science' in the English language.

16. *Three Dialogues between Hylas and Philonous*, Berkeley, pg. 73.

17. As above, pg. 73.

18. *An Essay Concerning Human Understanding*, Locke, pg. 135.

19. As above, pg. 143.

20. As above, pg. 144.

21. Formica was invented in 1913.

22. *Mysticism and Logic,* Russell, pg. 140.

23. *The Feeling of What Happens,* Damasio pg. 9.

Part II: From Colour to Consciousness

5. MEGA-MIND AND MICRO-MIND

1. *Perception*, B.K. Matilal.

2. As above, pg. 184.

3. As above, pg. 190.

4. As above, pg. 195.

6. MIND AND BODY

1. See 'Facing Up to the Problem of Consciousness,' David Chalmers in *Journal of Consciousness Studies 2,* 1995.

2. *Key Philosophical Writings: René Descartes,* Ed. E Chavez-Arvizo, pg. 181.

3. As above, pg. 181.

4. As above, pg. 147/148.

5. See 'Facing Up to the Problem of Consciousness', David Chalmers in *Journal of Consciousness Studies 2,* 1995.

6. As above.

7. As above.

8. 'The Puzzle of Conscious Experience', David Chalmers in *Scientific American* December 1995.

9. 'Facing Up to the Problem of Consciousness', David Chalmers in *Journal of Consciousness Studies 2,* 1995.

10. Even if the brain were understood as operating like a computer – as it does according to a doctrine called 'functionalism' – there would be discrete units in the form of its lowest level functions.

11. *Geons, Black Holes and Quantum Foam: A Life in Physics,* J.A. Wheeler, pg. 234.

12. See Nigel J.T. Thomas' article, 'Colour Realism: Towards a Solution to the "Hard Problem"', in *Consciousness and Cognition,* 2001.

13. As above.

Part III: Colours as Relations

7. Rescuing Direct Realism

1. 'Color Realism and Colour Science' by A. Byrne and D.R. Hilbert in *Behavioural and Brain Sciences 26,* 1: 3-21, 2003.

2. Technically this position might be referred to as 'chromatically-variable direct realism' but for simplicity I continue simply to call it 'direct realism', having already distinguished it from the chromatically-invariant form of 'naïve realism'.

8. Relational Colours: A New Entity

1. Both quoted in *The Home Planet,* Kelley, Ch. 2.

2. Also, intriguingly it has been suggested (e.g. Brower in his essay titled 'Abelard's Theory of Relations: Reductionism and the Aristotelian Tradition') that some thinkers in antiquity may have understood relations not in the modern fashion — as holding between entities — but rather as pointing from one entity to another. This is perhaps understandable given that they lived at a time when direct realism, and hence the (relational) form of sensory qualities as presenting or pointing outwardly across space, was largely unquestioned.

3. See for example 'Colour Properties and Colour Ascriptions: A Relationist Manifesto' by Jon Cohen, 2003. http://aardvark.ucsd.edu/color/relational.html

4. *Colour Vision*, Evan Thompson, pg.244.

5. As above, pg.244.

6. See Edward Averill's article, 'The Relational Nature of Colour' in *The Philosophical Review* July 1992.

9. Sense Qualities as Relations

1. For example, is the taste of mint not identical to the smell of mint? Or a coffee taste the same as a coffee smell?

2. This is suggested, if nothing else, by how widespread the sense of touch, or similar mechanical senses are amongst life-forms.

10. A Relational Theory of Sensory Qualities

1. Until depletion of energy means information content is lost.

11. Problems

1. *Three Dialogues between Hylas and Philonous*, Berkeley, pg. 73.

2. *Psychology: The Science of Behaviour,* Carlson and Buskist, pg. 118.

3. As above, pg. 589.

4. *An Essay Concerning Human Understanding* Locke, pg. 138.

5. *Supersense: Perception in the Animal World,* Downer, pg. 57.

6. As above, pg. 57.

7. As above, pg. 56.

8. *A Short History of the Universe,* Silk, pg. 120.

9. *Concepts of Modern Physics,* Beiser, pg. 123.

10. As above, pg. 479.

13. Color and Newton's Experiment

1. *Opticks,* Newton, pg. 124.

2. *Three Dialogues between Hylas and Philonous*, Berkeley, pg. 73.

14. The Bending of Light

1. Apart from the single refraction at the water-air interface giving rise to an apparent displacement in location, which for simplicity's sake I will ignore in the remainder of what follows.

15. A Bees' Eye View

1. A secondary problem raised by animal sensory consciousness is that it undermines an idea implicit to direct realism, namely that every object under constant lighting conditions has a single definite colour. If animals can see an object (or would if they were present) with a different colour than humans then what basis can there be to any claim that the colour apparent to humans is the 'actual' one? After all man is only another kind of primate and even amongst humans there is a certain amount of variation in colour experience due to the presence of degrees of colour blindness in the population.

2. Unlike the 'camera' eyes of many mammals – including man – which use a lens to bring a small patch in the centre of the field of view into sharp focus, the compound eye has no variable focusing capability. It makes up for this by being composed of many light-receiving units that cover the entire field of view. The result is that the animal can see with equal clarity, but some loss of resolution, over the entire field.

3. *Supersense: Perception in the Animal World,* Downer, pg. 57.

4. As above, pg. 57.

5. As above, pg. 57.

6. This stage in sensory processing is crucial because, as I previously suggested, and as we saw in more detail in a subsequent chapter, the primary contribution made by observers to the sensory process is that of the forming of such memories.

7. On the other hand it is a corollary of relational direct realism that a multiplicity of colour qualities exist in the external world where there may only seem to be one. As explained in the present chapter this is required in order to account for human-animal visual conflicts.

8. *An Essay Concerning Human Understanding,* Locke, pg. 138.

16. JEWEL IN THE SKY

1. *Psychology: The Science of Behaviour,* Carlson and Buskist, pg. 179.

2. As above, pg. 118.

3. As above, pg. 118.

4. *Some Must Watch Whilst Some Must Sleep*, Dement.

5. *Psychology: The Science of Behaviour,* Carlson and Buskist, pg. 118.

6. As above pg. 589.

17. THE COLOURS OF SPACE

1. *Astronomy: A Step By Step Guide,* Dunlop, pg. 14.

2. *The Concept of Nature,* Whitehead, pg. 21.

3. *From Atoms to Infinity,* Gribbin, pg. 167.

4. *A Short History of the Universe,* Silk pg. 68.

5. The almost infinitesimally small 'Planck' scale – at approximately twenty orders of magnitude smaller than the nucleus of an atom (see *Three Roads to Quantum Gravity,* Smolin, pg. 9) – provides the units of distance, time and energy at which quantum gravity effects may become significant (as above, pg. 217).

6. *Geons, Black Holes & Quantum Foam: A Life in Physics,* Wheeler, pg. 247.

7. *Three Roads to Quantum Gravity,* Smolin, pg. 106.

8. *The Fabric of The Cosmos,* Greene, pg. 332.

9. As above, pg. 490.

Part IV: Consciousness (Solving the Hard Problem)

18. THE VISUAL OPENING

1. *On Having No Head,* Harding, pg. 2.

19. SENSORY MEMORY

1. *Psychology: The Science Of Behaviour,* Carlson and Buskist, pg. 233.

2. *Introducing Psychological Research,* Banyard and Grayson, pg. 298.

3. *Memory: The Key To Consciousness,* Thompson and Madigan, pg. 28.

4. As above, pg. 27.

5. *Introducing Psychological Research,* Barnyard and Grayson, pg. 298.

6. *Psychology: The Science Of Behaviour,* Carlson and Buskist, pg. 233.

7. On this picture of man's experiential faculties as constituted by memory, the heightened awareness of visual forms know as 'gestalts' can be interpreted as the en-registration in memory of patterns detected in the incoming information stream by the processes of visual memory capture.

20. On Having No Mind

1. Any appearance that this sentence has of being self-contradictory should not be taken too seriously. My meaning could equally be conveyed by the words 'I suggest that minds do not exist'.

2. It is sometimes proposed that certain species of animal lack awareness of other individuals as a result of having no 'theory of mind'. But it could be that if animals lack anything it is a 'theory of perception' and hence they are untroubled by concepts of mind.

3. Although one may, of course, *conceive* of such qualities.

4. Rather than, as conventionally conceived, a simple sequential mode.

5. The biological locale of human memory has not yet been fully identified, but it is certainly not a prerequisite of memory that an organism has a sophisticated brain. Relatively simple animals such as slugs and worms display forms of behavioural memory. This raises the intriguing possibility that even in humans memory may be a process that is distributed throughout the whole-body, and occurring at the cellular level. If memory takes this form, and phantom-limbs arise as a result of cell-based memory processes regenerating previous information patterns, then the phenomenon may be felt by organisms without complex brains – such as worms. It is even conceivable that a plant such as a tree, despite having no central nervous system, may, if it possesses a cell-based form of memory capable of recreating pre-amputation patterns of information, feel a 'phantom-branch' when it suffers the loss of a physical branch.

21. On Having no Mind–body Problem

1. 'Facing Up to the Problem of Consciousness' in the *Journal of Consciousness*

Studies 2 (3):200-19, 1995, Chalmers.

2. As above.

22. THE HARD PROBLEM OF REALITY

1. For a discussion of this see Lockwood's essay 'The Grain Problem' in *Objections to Philosophy,* H. Robinson (ed.).

Part V: Consequences

23. A WORLD OF VALUE

1. Holt never explained how the sensory qualities that made up objects' appearances could exist in a material realm constituted of atoms and molecules, nor how the 'selection' process of perception worked.

2. It was possibly an inkling of this quandary which led Berkeley to his idea that material objects were perceived by God, as expressed in his notion that 'To be is to be perceived'.

3. I do not claim that it is in fact the case that the meaning of music resides in the external world. Only that the external existence of the sensory quality of sound makes this conceivable as a possibility.

4. Potentially strong grounds on which the prevailing account of perception may lie open to criticism.

5. In a letter to his patron, referred to in *Colour and Meaning,* Gage, pg. 146.

6. Adam Smith was a Scottish social thinker and economist. His seminal work *An Inquiry into the nature and causes of the Wealth of Nations* (1776) is generally regarded as having been one of the major influences on the early growth of '*laissez-faire*' capitalism.

24. THE FUNCTION OF THE NEURON

1. *Consciousness,* Carter, pg. 118.

2. *Consciousness: How Matter Becomes R. Imagination*, Edelman and Tononi.

3. As above, pg. 164–7

4. Part of the process that occurs in the perceptual system – quite possibly simultaneous with storage in primary memory must be for the information to be passed on to further areas of the brain for conversion into a behavioural response (and also eventual long-term memory storage).

5. See J. Kim's article 'Multiple Realization and the Metaphysics of Reduction' in T. O'Connor & D. Robb (eds.) *Philosophy of Mind Contemporary Readings*, 2003.

BIBLIOGRAPHY

Averill E. W. (1992), 'The Relational Nature of Color' in *The Philosophical Review*, 101, No. 3 (July 1992).

Banyard P. and Grayson A. (2000), *Introducing Psychological Research*, New York: Palgrave MacMillan.

Beiser A. (1995), *Concepts of Modern Physics,* New York: McGraw-Hill.

Berkeley G. (2005), *Three Dialogues Between Hylas and Philonous*, Oxford: Oxford University Press.

Brower J.E. (1998), 'Abelard's Theory of Relations: Reductionism and the Aristotelian Tradition' in *The Review of Metaphysics*, March 1st 1998, Washington: Catholic University of America.

Burns K. (2004), *Eastern Philosophy,* London: Arcturus.

Byrne A. and Hilbert D. R. (2003), 'Color Realism and Color Science' in *Behavioural and Brain Sciences* 26, 1: 3-21.

Carlson N.R. and Buskist W. (1997), *Psychology: The Science Of Behaviour,* Boston: Allyn and Bacon.

Carter R. (2002), *Consciousness,* London: Weidenfeld & Nicolson.

Chalmers D. (1995a), 'Facing Up to the Problem of Consciousness' in the *Journal of Consciousness Studies* 2(3): 200-19, 1995.

Chalmers D. (1995b), 'The Puzzle of Conscious Experience' in *Scientific American,* December 1995, pg. 62-68.

Cohen J. (2003), 'Color Properties and Color Ascriptions: A Relationalist Manifesto' - http://aardvark.ucsd.edu/color/relational.html

Damasio A. (2000), *The Feeling of What Happens,* London: Vintage.

Dement W. C. (1974), *Some Must Watch While Some Must Sleep,* San Francisco: W.H. Freeman.

Descartes R. (1997), *Key Philosophical Writings* (Ed. By E. Chavez-Arvizo), Ware: Wordsworth Editions Ltd.

Descartes R. (1998), *The World and Other Writings,* Cambridge: Cambridge University Press.

Descartes R. (1999), *Meditations,* London: Penguin.

Downer J. (1988), *Supersense: Perception in the Animal World,* London: BBC Books.

Dunlop, S. (1985), *Astronomy: A Step by Step Guide to the Night Sky,* Feltham: Hamlyn.

Edelman G.M. and Tononi G. (2001) *Consciousness How Matter Becomes R. Imagination,* London: Penguin Books.

Gage J. (2003), *Colour and Meaning,* London: Thames & Hudson.

Grayling A.C. (2005), *Descartes,* London: The Free Press.

Greene B. (2005), *The Fabric of The Cosmos,* London: Penguin Books.

Gribbin J. and M. (2006), *From Atoms to Infinity,* Cambridge: Icon Books.

Harding D. (1986), *On Having No Head,* London: Arkana.

Kelley K. (1988), *The Home Planet,* Massachusetts: De Capo Press

Kim J. (2003), 'Multiple Realization and the Metaphysics of Reduction' in T. O'Connor & D. Robb (eds) *Philosophy of Mind: Contemporary Readings* (2003), London: Routledge.

Kosslyn S.M. (1999), *Image and Brain,* Massachusetts: The MIT Press.

Locke J. (2004), *An Essay Concerning Human Understanding,* London: Penguin Books.

Lockwood M. (1993), 'The Grain Problem' in H. Robinson (ed.) *Objections to Physicalism,* Oxford: Clarendon Press

Lowe E. J. (2005), Locke, London & New York: Routledge.

Matilal B.K. (1986), *Perception,* Oxford: Oxford University Press.

Minnaert M. (1954), *Light & Colour in the Open Air,* New York: Dover Publications.

Newton I. (1952/1730), *Opticks, Or a Treatise of the Reflections, Refractions, Inflections & Colours of Light,* New York: Dover Publications.

Newton I. (1953/1671), 'The new theory about light and colors' in H.S. Thayer (ed.) *Newton's Philosophy of Nature: Selections from his Writings,* New York and London: Hafner Publishing Company.

Russell B. (1994), *Mysticism And Logic,* London: Routledge.

Sabra A.I. (1967), *Theories of Light From Descartes to Newton,* London: Oldbourne Book Co.

Silk J. (1994), *A Short History of The Universe,* New York: Scientific American Library.

Smolin L. (2001), *Three Roads to Quantum Gravity,* London: Orion Books.

Thomas N.J.T. (2001), *Color Realism: Toward a Solution to the 'Hard Problem'* *in Consciousness and Cognition*, 10, 140–145 (2001), Academic Press.

Thompson E. (1995), *Colour Vision*, London: Routledge.

Thompson R. and Madigan S. (2005), *Memory: The Key To Consciousness,* Washington: Joseph Henry Press.

Tootell, R.B.H., M.S. Silverman, E. Switkes, R.L. De Valois. (1982) 'Deoxyglucose analysis of retinotopic organisation in primate striate cortex.' *Science* 218: 902–904.

Watson R. (2002), *Cogito Ergo Sum: The Life of René Descartes*, Boston: David R. Godine.

Westfall R.S. (1980), *Never At Rest: A Biography of Isaac Newton*, Cambridge: Cambridge University Press.

Wheeler, J.A. (1998), *Geons, Black Holes & Quantum Foam, A Life in Physics,* New York: W.W. Norton & Co.

Whitehead, A.N., (1978), *The Concept of Nature,* Cambridge: Cambridge University Press.

Wiener, Philip P. (ed) (1973), *Dictionary of the History of Ideas*, *Studies of Selected Pivotal Ideas,* New York: Charles Scribner's Sons.

Wittgenstein L. (1922/1974), *Tractatus Logico-Philosophicus,* London: Routledge.

Zajonc A. (1993) *Catching The Light: The Entwined History of Light and Mind,* London: Bantam Press.

ACKNOWLEDGEMENTS

There are a number of people without whose input this book would never have seen the light of day. So I would particularly like to thank (in chronological order):

My father and mother, who encouraged a questioning attitude in me (as in all of their children) from an early age. Oliver Ramsbotham (Marlborough); his pioneering teaching of philosophy during 'religious education' lessons inspired me, and I am sure many of my contemporaries, to become interested in the subject. David Milligan (Bristol) who gave early encouragement to my first tentative under-graduate attempts at creative thinking. Jon Smith; it was during long, drug-fuelled discussions with my fellow student Jon that the seed ideas first appeared which eventually grew, many years later, into those presented here. The staff of HITTU (Frenchay Hospital, Bristol) who saved my life in the aftermath of a critical illness. Rebecca Gillieron and Catheryn Kilgarriff (Marion Boyars Publishers) who took seriously a proposal by an obscure writer and commissioned it for publication.

In addition I would also like to thank: Andrew Woodfield for his friendship and assistance on philosophical matters over many years. The Neurology department at Frenchay Hospital. Rebecca Gillieron, whose input as editor has greatly improved the readability and quality of the present text. My wife Fiona and children Indigo, Saffron and Kir who have stoically put up with my involvement in this project over many years.